Lecture Notes in Mathematics

Volume 2286

This series reports on new developments in all areas of mathematics and their applications - quickly, informally and at a high level. Mathematical texts analysing new developments in modelling and numerical simulation are welcome. The type of material considered for publication includes:

1. Research monographs
2. Lectures on a new field or presentations of a new angle in a classical field
3. Summer schools and intensive courses on topics of current research.

Texts which are out of print but still in demand may also be considered if they fall within these categories. The timeliness of a manuscript is sometimes more important than its form, which may be preliminary or tentative.

Titles from this series are indexed by Scopus, Web of Science, Mathematical Reviews, and zbMATH.

More information about this series at http://www.springer.com/series/304

Scott Harper

The Spread of Almost Simple Classical Groups

Springer

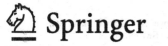

Scott Harper
School of Mathematics
University of Bristol
Bristol, UK

ISSN 0075-8434 ISSN 1617-9692 (electronic)
Lecture Notes in Mathematics
ISBN 978-3-030-74099-3 ISBN 978-3-030-74100-6 (eBook)
https://doi.org/10.1007/978-3-030-74100-6

Mathematics Subject Classification: Primary: 20E32, 20F05; Secondary: 20E28, 20P05

This Springer imprint is published by the registered company Springer Nature Switzerland AG.
The registered company address is: Gewerbestrasse 11, 6330 Cham, Switzerland

Preface

Group generation is a classical and fascinating area of mathematical research. The finite simple groups are the building blocks of all finite groups, and, by a theorem of Steinberg, every finite simple group can be generated by two elements. In 2000, Guralnick and Kantor resolved a 1962 question of Steinberg by proving that in a finite simple group every nontrivial element belongs to a generating pair. Groups with this property are said to be $\frac{3}{2}$-generated.

Which finite groups are $\frac{3}{2}$-generated? It is easy to see that every proper quotient of a $\frac{3}{2}$-generated group is cyclic, and in 2008, Breuer, Guralnick and Kantor made the striking conjecture that this condition alone provides a complete characterisation of the finite groups with this property. This conjecture has recently been reduced to the almost simple groups, and results of Piccard (1939) and Woldar (1994) show that the conjecture is true for almost simple groups whose socles are alternating or sporadic groups. Therefore, the central focus is now on the almost simple groups of Lie type.

In this monograph, we prove a strong version of this conjecture for almost simple classical groups, building on earlier work of Burness and Guest (2013) and the author (2017). More precisely, the uniform spread of a finite group G is the greatest k for which there exists a conjugacy class C such that for any nontrivial elements $x_1, \dots, x_k \in G$ there exists $y \in C$ satisfying $\langle x_i, y \rangle = G$ for all i. We show that for every almost simple classical group G for which $G/\operatorname{soc}(G)$ is cyclic, the uniform spread of G is at least two, unless it is isomorphic to the symmetric group of degree six, which has uniform spread zero (but is still $\frac{3}{2}$-generated). We also prove that the uniform spread of these groups tends to infinity if the size of the underlying field tends to infinity.

To prove these results, we utilise a probabilistic approach introduced by Guralnick and Kantor. This requires a detailed analysis of automorphisms, fixed point ratios and subgroup structure of almost simple classical groups, so the first half of this monograph is dedicated to these general topics. In particular, we give a general exposition of the useful technique of Shintani descent, which plays an important role throughout.

Therefore, this monograph has two purposes. On the one hand, for a broad audience, including graduate students new to the area, it provides an introduction to the almost simple classical groups and some powerful techniques for working with them. On the other hand, for experts in the field, it provides a detailed analysis of the spread of these groups, as part of a wider current research project.

This monograph has its origins in the author's PhD research at the University of Bristol, where he developed a taste for group generation. He gratefully acknowledges the financial support of the Engineering and Physical Sciences Research Council and the Heilbronn Institute for Mathematical Research during his PhD studies. The author sincerely thanks his PhD supervisor Dr. Tim Burness for introducing him to this fascinating subject and for his generous support over the course of this work. He wishes to thank Professor Robert Guralnick for his encouragement and Professors Chris Parker and Jeremy Rickard for discussions about this work and reading earlier versions of this monograph. Finally, he thanks the anonymous referees for their careful reading of this monograph and their helpful comments.

Bristol, UK Scott Harper

Contents

Chapter 1
Introduction

The topic of generating sets for groups has a history dating back to the earliest days of group theory, and it has led to a broad and rich literature, especially in the context of finite simple groups. In 1962, by exhibiting explicit pairs of generators, Steinberg [61] proved that every finite simple group of Lie type is 2-generated. In light of the Classification of Finite Simple Groups, together with results on alternating and sporadic groups [3], we now know every finite simple group is 2-generated.

In the opening of his 1962 paper, Steinberg writes

> It is possible that one of the generators can be chosen of order 2, as is the case for the projective unimodular group, or even that one of the generators can be chosen as an arbitrary element other than the identity, as is the case for the alternating groups. Either of these results, if true, would quite likely require methods much more detailed than those used here.

This motivates the following definition, which is central to this monograph.

Definition A group G is $\frac{3}{2}$*-generated* if for every nontrivial element $g \in G$, there exists an element $h \in G$ such that $\langle g, h \rangle = G$.

In recent years, probabilistic methods have been very successful in solving many formidable deterministic problems in group theory (see, for example, [18, 53, 57]). Indeed, through a probabilistic approach, Guralnick and Kantor [38] proved that every finite simple group is $\frac{3}{2}$-generated, resolving the above question of Steinberg.

Classifying the 1-generated groups is trivial and classifying the 2-generated groups is impossible. Can we classify the $\frac{3}{2}$-generated groups? It is straightforward to demonstrate that every proper quotient of an arbitrary $\frac{3}{2}$-generated group is necessarily cyclic. In 2008, Breuer, Guralnick and Kantor [11] conjectured that this evidently necessary condition is actually sufficient for finite groups.

Conjecture ($\frac{3}{2}$-Generation Conjecture) *A finite group is $\frac{3}{2}$-generated if and only if every proper quotient is cyclic.*

S. Harper, *The Spread of Almost Simple Classical Groups*, Lecture Notes in Mathematics 2286, https://doi.org/10.1007/978-3-030-74100-6_1

Note that this necessary condition for $\frac{3}{2}$-generation is not sufficient for infinite groups; for example, the alternating group A_∞ is simple but not finitely generated, let alone $\frac{3}{2}$-generated. However, in [30], Donoven and the author proved that two natural families of infinite groups generalising Thompson's group V are $\frac{3}{2}$-generated, thus providing the first known examples of infinite $\frac{3}{2}$-generated groups (other than the infinite cyclic group and Tarski monsters).

The $\frac{3}{2}$-Generation Conjecture is true for soluble groups [9, Theorem 2.01], and for insoluble groups the conjecture has recently been reduced to the almost simple groups [22]. Therefore, to prove the $\frac{3}{2}$-Generation Conjecture it is enough to prove that $\langle T, \theta\rangle$ is $\frac{3}{2}$-generated for all nonabelian finite simple groups T and all automorphisms $\theta \in \mathrm{Aut}(T)$.

The alternating and symmetric groups of degree at least 5 have been known to be $\frac{3}{2}$-generated since the work of Piccard in 1939 [56], to which Steinberg refers in the quotation above. In addition, the $\frac{3}{2}$-generation of the relevant almost simple sporadic groups (and the two further almost simple cyclic extensions of A_6) follows from the computational results of Breuer, Guralnick and Kantor [11] (see also [62]). Therefore, to prove the $\frac{3}{2}$-Generation Conjecture, it suffices to focus on almost simple groups of Lie type. In this monograph, we prove the $\frac{3}{2}$-Generation Conjecture for almost simple classical groups. The exceptional groups pose different challenges and this is the topic of a forthcoming paper [22].

Theorem 1 *Let G be an almost simple classical group. Then G is $\frac{3}{2}$-generated if every proper quotient of G is cyclic.*

We actually prove a much stronger version of this theorem. To state our main results we must introduce some natural generalisations of $\frac{3}{2}$-generation.

Definition Let G be a finite noncyclic group.

(i) The *spread* of G, written $s(G)$, is the greatest k such that for any k nontrivial elements $x_1, \dots, x_k$, there exists $y \in G$ such that

$$\langle x_1, y\rangle = \langle x_2, y\rangle = \cdots = \langle x_k, y\rangle = G.$$

(ii) The *uniform spread* of G, written $u(G)$, is the greatest k for which there exists a fixed conjugacy class C such that for any k nontrivial elements $x_1, \dots, x_k$, there exists an element $y \in C$ satisfying the above equalities.

Observe that $s(G) \geqslant u(G)$ and that $s(G) \geqslant 1$ if and only if G is $\frac{3}{2}$-generated, so these invariants extend the idea of $\frac{3}{2}$-generation. If G is simple, then Breuer, Guralnick and Kantor [11] proved that $u(G) \geqslant 2$ with equality if and only if $G \in \{A_5, A_6, \Omega_8^+(2)\}$ or G is $\mathrm{Sp}_{2m}(2)$ for $m \geqslant 3$. This generalises the fact that $s(G) \geqslant 1$ for simple groups G. In addition, Guralnick and Kantor [38] proved that if (G_i) is a sequence of simple groups of Lie type where G_i is defined over $\mathbb{F}_{q_i}$, then $u(G_i) \to \infty$ if $q_i \to \infty$. Later Guralnick and Shalev [42] determined exactly when $|G_i| \to \infty$ but $u(G_i)$ is bounded.

We now present our main results on uniform spread. Here, and throughout, $\operatorname{soc}(G)$ denotes the socle of G.

Theorem 2 *Let G be an almost simple classical group such that $G/\operatorname{soc}(G)$ is cyclic. Then $u(G) \geqslant 2$, unless $G \cong S_6$, in which case $u(G) = 0$ and $s(G) = 2$.*

Theorem 3 *Let (G_i) be a sequence of almost simple classical groups where G_i is defined over $\mathbb{F}_{q_i}$ and $G_i/\operatorname{soc}(G_i)$ is cyclic. Then $u(G_i) \to \infty$ as $q_i \to \infty$.*

In 2013, Burness and Guest [21] proved Theorems 2 and 3 for almost simple groups with socle $\mathrm{PSL}_n(q)$. They followed the probabilistic approach of Guralnick and Kantor in [38] but brought a powerful new technique to the problem: *Shintani descent* (see p. 4). Inspired by this work, the author proved Theorems 2 and 3 for symplectic and odd-dimensional orthogonal groups in [43] using similar methods.

However, as we explain below, the methods used in these previous papers are not enough to handle the remaining classical groups, which present new challenges. This monograph addresses these challenges and completes the proof of Theorems 2 and 3 by proving the following two results.

Theorem 4 *Let G be an almost simple group with socle $\mathrm{P\Omega}^{\pm}_n(q)$ (n even) or $\mathrm{PSU}_n(q)$ such that $G/\operatorname{soc}(G)$ is cyclic. Then $u(G) \geqslant 2$, unless $G \cong S_6$.*

Theorem 5 *Let (G_i) be a sequence of almost simple groups, where G_i has socle $\mathrm{P\Omega}^{\pm}_{n_i}(q_i)$ (n_i even) or $\mathrm{PSU}_{n_i}(q_i)$ and $G_i/\operatorname{soc}(G_i)$ is cyclic. Then $u(G_i) \to \infty$ as $q_i \to \infty$.*

Therefore, in this monograph, we concentrate on even-dimensional orthogonal groups and unitary groups, with the aim of proving Theorems 4 and 5. One case when $\operatorname{soc}(G) = \mathrm{PSL}_n(q)$ was omitted in [21], so we also prove Theorems 2 and 3 in this special case (see Remark 6.2.8).

Let us make some remarks on the statements of the main theorems.

Remark 1.1 Let G be a finite group. The $\frac{3}{2}$-Generation Conjecture avers that $s(G) \geqslant 1$ if every proper quotient of G is cyclic. A stronger version of this conjecture is that $s(G) \geqslant 2$ if every proper quotient of G is cyclic, and this would imply that there do not exist any finite groups with $s(G) = 1$ (see [19, Conjecture 3.16]). Theorem 2 proves this stronger conjecture for almost simple classical groups.

Remark 1.2 Let (G_i) be a sequence of almost simple classical groups where $\operatorname{soc}(G_i)$ has natural module $\mathbb{F}_{q_i}^{n_i}$ and $G_i/\operatorname{soc}(G_i)$ is cyclic. By Theorem 3, $u(G_i) \to \infty$ if $q_i \to \infty$, but it is difficult to determine when $n_i \to \infty$ implies $u(G_i) \to \infty$ for bounded q_i. Even for simple groups, $G_i = \mathrm{Sp}_{2i+2}(2)$ gives an example where $n_i \to \infty$ but $u(G_i) = 2$ for all i (see [42, Proposition 2.5]). Almost simple groups provide an even greater challenge: for instance, if $G_i = \mathrm{Aut}(\mathrm{PSL}_{2i+1}(2)) = \mathrm{PSL}_{2i+1}(2).2$, then $u(\operatorname{soc}(G_i)) \to \infty$ but $u(G_i) \leqslant 8$ for all i (see [21, Theorem 4]). Determining when the uniform spread of almost simple classical groups is bounded will feature in future work [44].

Remark 1.3 Let us note that determining the exact spread and uniform spread of a finite group is a difficult problem. Notably, the spread of odd-degree alternating groups is not known in general, but it can be arbitrarily large (see [24, Remark 3] and the references therein); however, by [9, (3.01)–(3.05)], it is known that $s(A_n) = 4$ if $n \geqslant 8$ is even. In addition, $s(\mathrm{PSL}_2(q))$ is not known when $q \equiv 3 \pmod 4$ (see the discussion in [24, Remark 5]), and for sporadic groups the exact spread is only known in two cases ($s(\mathrm{M}_{11}) = 3$ and $s(\mathrm{M}_{23}) = 8064$, see [33]).

Remark 1.4 We have already observed that $s(S_6) = 2$ but $u(S_6) = 0$. It is worth noting that $s(G)$ and $u(G)$ can be different positive integers. For example, if $n \geqslant 5$ is odd, then $s(S_n) = 3$ and $u(S_n) = 2$ (see [24, Theorem 3] and the references in [24, Remark 3]). However, the only known family of nonabelian simple groups for which $s(G) - u(G)$ is unbounded is $G = \mathrm{PSL}_2(p)$ where p is a prime number satisfying $p \equiv 3 \pmod 4$ (see [24, Proposition 7.4]).

We now turn to a brief discussion of the techniques employed in this monograph; the opening of Chap. 5 gives a more technical account of the particular challenges that we have to overcome. For this discussion, G is an almost simple classical group such that $G/\mathrm{soc}(G)$ is cyclic.

The framework for proving Theorems 4 and 5 is given by the probabilistic method introduced by Guralnick and Kantor [38] (see Sect. 2.1). The general idea is to select an element $s \in G$ and show that s^G witnesses $u(G) \geqslant k$. To do this, we let $P(x, s)$ be the probability that $\langle x, z \rangle \neq G$ for a random conjugate z of s. Evidently, $u(G) \geqslant 1$ if $P(x, s) < 1$ for all nontrivial $x \in G$. Indeed, $u(G) \geqslant k$ if $P(x, s) < \frac{1}{k}$ for all prime order $x \in G$ (see Lemma 2.1.1).

Let $\mathcal{M}(G, s)$ be the set of maximal subgroups of G that contain s. In addition, for $H \leqslant G$ and $x \in G$, let $\mathrm{fpr}(x, G/H)$ be the *fixed point ratio* of x in the action of G on G/H. We will see in Lemma 2.1.1 that

$$P(x, s) \leqslant \sum_{H \in \mathcal{M}(G,s)} \mathrm{fpr}(x, G/H).$$

Therefore, our probabilistic method has three steps: select an appropriate element $s \in G$, determine $\mathcal{M}(G, s)$ and use fixed point ratio estimates to bound $P(x, s)$.

Selecting a viable element $s \in G$ is perhaps the most interesting and challenging aspect of the proofs. Write $G = \langle T, \theta \rangle$ where $T = \mathrm{soc}(G)$ and $\theta \in \mathrm{Aut}(T)$. If s^G witnesses $u(G) \geqslant k > 0$, then s is not contained in any proper normal subgroup of G, so we may assume that $s \in T\theta$. Consequently, we need to understand the conjugacy classes in the coset $T\theta$.

We view the finite groups of Lie type as the fixed points under Steinberg endomorphisms of simple algebraic groups, and this perspective allows us to exploit Shintani descent [45, 58], a technique which has seen great utility in the character theory of almost simple groups of Lie type (see [26, 29, 46, 59] for example). At the heart of this method is a bijection with useful group theoretic properties that, given a connected algebraic group X, a Steinberg endomorphism σ of X and an integer $e > 1$, provides a correspondence between the conjugacy classes of elements in the coset $X_{\sigma^e}\sigma$ and in the subgroup X_σ (here X_σ is the set of fixed points of σ in X). In essence, we use this bijection to transform a problem about almost simple groups into one about simple groups.

Shintani descent was used by Burness and Guest in [21] in the context of linear groups, and this technique was extended in [43] to overcome various difficulties and subtleties that the symplectic groups posed (such as the disconnected orthogonal subgroups in even characteristic and the presence of graph-field automorphism of $\mathrm{Sp}_4(2^f)$). In this monograph we present a general formalism of Shintani descent for applications to all almost simple groups of Lie type, which we anticipate will be useful more generally. Shintani descent is introduced in Chap. 3, where we provide crucial new results that allow us to handle the novel challenges posed by the twisted minus-type orthogonal and unitary groups.

Our framework for understanding $\mathcal{M}(G, s)$ is provided by Aschbacher's subgroup structure theorem for finite classical groups [1], which asserts that the maximal subgroups of classical groups are either the stabilisers of geometric structures on the natural module or they arise from an absolutely irreducible representation of a quasisimple group. By studying how a suitable power of our chosen element acts on the natural module, we can constrain the maximal subgroups that could contain this element.

The common strategy of choosing s to have a large and restrictive order cannot typically be employed for this problem, so we require different techniques. This obstacle occurs because the element s is contained in the nontrivial coset $T\theta$ and in many cases this forces s to have a comparatively small order (indeed, even determining the possible element orders in this coset is nontrivial). In Remark 5.4.9, we use Shintani descent to explain this issue more precisely.

Once we have a description of $\mathcal{M}(G, s)$, we use fixed point ratio estimates to bound $P(x, s)$. There is an extensive literature on fixed point ratios for primitive actions of almost simple groups, and these quantities have found applications to a vast range of problems, including the resolution of the Cameron–Kantor conjecture on base sizes of permutation groups [55] and the Guralnick–Thompson conjecture on monodromy groups [35]. In Chap. 4 we review some general results in this area, in particular the essentially best possible bounds on the fixed point ratios of nonsubspace actions of almost simple groups given in [14–17], which play a key role. In the same chapter, we prove some new fixed point ratio bounds that we require for our proofs; these bounds may be of independent interest.

Let us now highlight a combinatorial connection to this work. The *generating graph* of a group G is the graph $\Gamma(G)$ whose vertices are the nontrivial elements of G and where two vertices g and h are adjacent if $\langle g, h\rangle = G$. The generating graphs of the dihedral group D_8 and the alternating group A_4 are given in Fig. 1.1

Evidently, $\Gamma(G)$ has no isolated vertices if and only if G is $\frac{3}{2}$-generated. This dichotomy is demonstrated by $\Gamma(D_8)$ and $\Gamma(A_4)$, where we note that D_8 has a noncyclic quotient whereas A_4 does not. Further, if $s(G) \geqslant 2$, then $\Gamma(G)$ is connected with diameter at most 2. Therefore, the stronger version of the $\frac{3}{2}$-Generation Conjecture in Remark 1.1 has the striking interpretation that the generating graph of a finite group either has an isolated vertex or it is connected

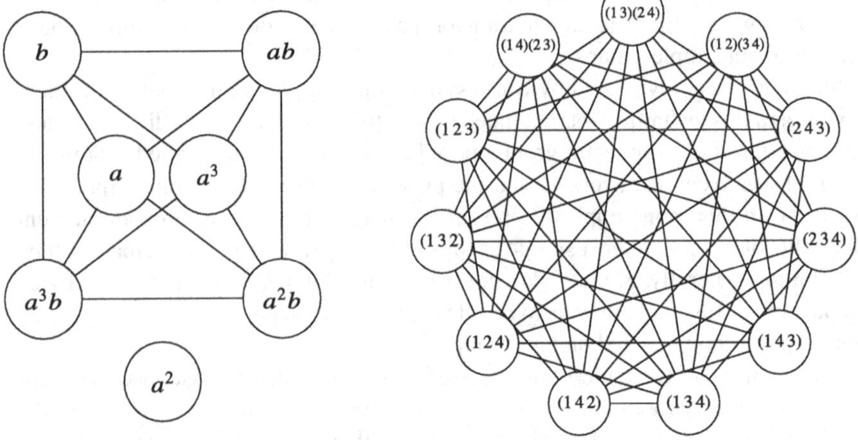

Fig. 1.1 The generating graphs of D_8 and A_4

with diameter at most two. By [11, Theorem 1.2], the diameter of the generating graph of any nonabelian finite simple group is two, and Theorem 2 implies that the same conclusion holds for almost simple classical groups G such that $G/\operatorname{soc}(G)$ is cyclic.

Many other natural questions about generating graphs have been investigated in recent years. For instance, if G is a sufficiently large simple group, then $\Gamma(G)$ is *Hamiltonian* (that is, has a cycle containing every vertex exactly once) [12]. Moreover, if $n \geqslant 120$, then the generating graphs $\Gamma(A_n)$ and $\Gamma(S_n)$ are Hamiltonian [32]. Indeed, it is conjectured that for all finite simple groups G of order at least four, the generating graph $\Gamma(G)$ is Hamiltonian if and only if every proper quotient of G is cyclic, which is another strengthening of the $\frac{3}{2}$-Generation Conjecture.

In a different direction, the *total domination number* of a graph Γ is the minimal size of a set S of vertices of Γ such that every vertex of Γ is adjacent to a vertex in S. In recent work of Burness and the author [23, 24], close to best possible bounds on the total domination number of generating graphs of simple groups were obtained, together with related probabilities. For instance, there are infinitely many finite simple groups G for which the total domination number of $\Gamma(G)$ is the minimal possible value of two (for example, A_p when $p \geqslant 13$ is prime, $\mathrm{PSL}_n(q)$ when $n > 3$ is odd, $E_8(q)$ and the Monster). This is a vast generalisation of the fact that these groups are $\frac{3}{2}$-generated.

For further reading on group generation, especially in the context of simple groups and probabilistic methods, see Burness' recent survey article [19]. The recent paper of Burness and the author [24] also features a detailed account of the spread of simple groups and related finite groups.

We conclude the introduction with an outline of the structure of this monograph. Chapter 2 introduces the almost simple classical groups, their subgroups, the formed spaces they naturally act on and their connection with simple algebraic groups. As noted above, in Chap. 3 we turn to Shintani descent, where we unify existing results in this area and provide new methods that allow us to handle all almost simple classical groups. Chapter 4 is dedicated to establishing bounds on fixed point ratios. In the opening sections of Chaps. 5 and 6, we study automorphisms and special elements of classical groups. Following this general exposition on almost simple classical groups, which we anticipate will be of wider interest, in the final three sections of Chaps. 5 and 6, we focus on the proofs of our main results on uniform spread.

Chapter 2
Preliminaries

In this chapter we introduce the requisite background material from the literature on almost simple classical groups, and we will use this opportunity to establish the notation that we use in this monograph.

Notational Conventions

Let a, b, n be positive integers and let G, H be groups. Throughout we write

(a, b) for the greatest common divisor of a and b
a_b for the greatest power of b dividing a
δ_{ab} for the Kronecker delta
$\log a$ for the *base two* logarithm of a
C_n (or simply n) for the cyclic group of order n
$G.H$ for an unspecified extension of G by H (with quotient H)
$G{:}H$ for an unspecified split extension of G by H

Groups always act on the right. Accordingly, matrices act on the right of row vectors, x^g denotes $g^{-1}xg$ and G/H is the set of right cosets of H in G.

2.1 Probabilistic Method

Probabilistic methods featuring fixed point ratios, introduced below, are a fruitful means of studying a wide range of problems, and the survey article [18] provides an excellent overview of this topic. In this section, we outline the probabilistic method for studying uniform spread introduced by Guralnick and Kantor [38].

S. Harper, *The Spread of Almost Simple Classical Groups*, Lecture Notes in Mathematics 2286, https://doi.org/10.1007/978-3-030-74100-6_2

Let G be a finite group acting on a finite set Ω. The *fixed point ratio* of $x \in G$ is

$$\mathrm{fpr}(x,\Omega) = \frac{\mathrm{fix}(x,\Omega)}{|\Omega|} \quad \text{where} \quad \mathrm{fix}(x,\Omega) = |\{\omega \in \Omega \mid \omega x = \omega\}|.$$

If $H \leqslant G$, then G acts transitively on G/H and one sees that

$$\mathrm{fpr}(x, G/H) = \frac{|x^G \cap H|}{|x^G|}.$$

We discuss recent work on fixed point ratios, particularly in the context of primitive actions of almost simple groups, at the opening of Chap. 4.

We outline the probabilistic method for studying uniform spread. For $x, s \in G$

$$P(x,s) = \frac{|\{z \in s^G \mid \langle x, z\rangle \neq G\}|}{|s^G|} \tag{2.1}$$

is the probability that x does not generate G with a (uniformly) randomly chosen conjugate of s. Let $\mathcal{M}(G,s)$ be the set of maximal subgroups of G that contain s. The following encapsulates the method (see [21, Lemmas 2.1 and 2.2]).

Lemma 2.1.1 *Let G be a finite group and let $s \in G$.*

(i) *For $x \in G$,*

$$P(x,s) \leqslant \sum_{H \in \mathcal{M}(G,s)} \mathrm{fpr}(x, G/H).$$

(ii) *If for all k-tuples $(x_1, \dots, x_k)$ of prime order elements of G*

$$\sum_{i=1}^{k} P(x_i, s) < 1,$$

then $u(G) \geqslant k$ with respect to the conjugacy class s^G.

We conclude this section with an elementary observation.

Lemma 2.1.2 *Let G be a finite group, let $H \leqslant G$ and let $x \in G$. Then the number of G-conjugates of H that contain x is* $\mathrm{fpr}(x, G/H) \cdot |G : N_G(H)|$.

2.2 Classical Groups

Let F be a finite or algebraically closed field of characteristic $p > 0$ and $V = F^n$. Our notation for classical groups is mainly standard, but there is variation in the literature, so we will briefly outline the notation we use. For further background on classical groups see [2, Chapter 7], [20, Chapter 2] and [50, Chapter 2].

If κ is a bilinear, sesquilinear or quadratic form on V, then write $\mathrm{Isom}(V,\kappa)$, $\mathrm{Sim}(V,\kappa)$ and $\mathrm{Semi}(V,\kappa)$ for the groups of isometries, similarities and semisimilarities of κ, and write $\mathrm{SIsom}(V,\kappa)$ for $\mathrm{Isom}(V,\kappa) \cap \mathrm{SL}(V)$. Using this notation, Fig. 2.1 gives our notation for the classical groups. For projective groups, we adopt the standard convention that for $G \leqslant \Gamma\mathrm{L}(V)$ we write $\mathrm{P}G = GZ(V)/Z(V) \cong G/(G \cap Z(V))$, where $Z(V) = Z(\mathrm{GL}(V)) \leqslant \Gamma\mathrm{L}(V)$ is the group of scalar transformations on V.

Remark 2.2.1 Let us remark on the notation introduced in Fig. 2.1.

(i) *Unitary groups.* By κ being unitary, we require that F be a quadratic extension of F_0 with $\mathrm{Gal}(F/F_0) = \langle\sigma\rangle$, and we mean that κ is a nondegenerate σ-conjugate symmetric sesquilinear form on V. The notation $\Delta\mathrm{U}_n(F_0)$ is uncommon but follows [20]; we shall rarely need to refer to this group since $\mathrm{P}\Delta\mathrm{U}_n(F_0) = \mathrm{PGU}_n(F_0)$. We write $\mathrm{GL}^+ = \mathrm{GL}$ and $\mathrm{GL}^- = \mathrm{GU}$.

(ii) *Symplectic groups.* By κ being symplectic, we require that n be even, and we mean that κ is a nondegenerate alternating bilinear form on V.

(iii) *Orthogonal groups.* Let κ be a nondegenerate (n.d.) quadratic form Q with associated bilinear form $(\cdot,\cdot)$ defined as $(u, v) = Q(u+v) - Q(u) - Q(v)$. The *norm* of a vector $v \in V$ is (v, v).

(a) If $n = 2m$ is even, then there are at most two isometry types of nondegenerate quadratic forms Q, distinguished by their *Witt index*, the dimension of a maximal totally singular subspace. More precisely, if $F = \bar{F}$, then there is a unique type. If $F = \mathbb{F}_q$, there there are exactly two types, plus and minus, and we write $\mathrm{sgn}(Q) = \varepsilon$ when Q is ε-type. If q is odd, then these types are distinguished by the *discriminant* $D(Q) \in \mathbb{F}_q/(\mathbb{F}_q)^2 = \{\square, \boxtimes\}$:

$$D(Q) = \square \iff q^m \equiv \mathrm{sgn}(Q) \pmod 4, \tag{2.2}$$

interpreting $\mathrm{sgn}(Q)$ as 1 or -1 (see [50, Proposition 2.5.10]).

κ		$\mathrm{SIsom}(V,\kappa)$	$\mathrm{Isom}(V,\kappa)$	$\mathrm{Sim}(V,\kappa)$	$\mathrm{Semi}(V,\kappa)$
zero	$\mathrm{SL}_n(F)$	$\mathrm{SL}_n(F)$	$\mathrm{GL}_n(F)$	$\mathrm{GL}_n(F)$	$\Gamma\mathrm{L}_n(F)$
unitary	$\mathrm{SU}_n(F_0)$	$\mathrm{SU}_n(F_0)$	$\mathrm{GU}_n(F_0)$	$\Delta\mathrm{U}_n(F_0)$	$\Gamma\mathrm{U}_n(F_0)$
symplectic	$\mathrm{Sp}_n(F)$	$\mathrm{Sp}_n(F)$	$\mathrm{Sp}_n(F)$	$\mathrm{GSp}_n(F)$	$\Gamma\mathrm{Sp}_n(F)$
n.d. quadratic	$\Omega_n^\varepsilon(F)$	$\mathrm{SO}_n^\varepsilon(F)$	$\mathrm{O}_n^\varepsilon(F)$	$\mathrm{GO}_n^\varepsilon(F)$	$\Gamma\mathrm{O}_n^\varepsilon(F)$

Note: see Remark 2.2.1 for a description of F_0 and ε

Fig. 2.1 Notation for classical groups

(b) If n is odd, then V admits a nondegenerate quadratic form if and only if p is odd, in which case there is a unique similarity type of form and we write $\mathrm{sgn}(Q) = \circ$. If $F = \mathbb{F}_q$ for odd q, then there are two isometry types of nondegenerate quadratic form Q, distinguished by the discriminant $D(Q)$.

(iv) *The group* $\Omega_n^\varepsilon(F)$. Let Q be a nondegenerate quadratic form and assume $(n, F, \mathrm{sgn}(Q)) \neq (4, \mathbb{F}_2, +)$ (see [50, Proposition 2.5.9] in this case). If $p = 2$, then every element of $\mathrm{SO}_n^\varepsilon(F) = \mathrm{O}_n^\varepsilon(F)$ is a product of reflections and we define $\Omega_n^\varepsilon(F)$ as the group of elements that are a product of an even number of reflections (see [2, 22.7–22.9]). If p is odd, then $\mathrm{SO}_n^\varepsilon(F)$ is the group of elements that are a product of an even number of reflections and we define $\Omega_n^\varepsilon(F)$ as the kernel of the *spinor norm* $\mathrm{SO}_n^\varepsilon(F) \to F^\times/(F^\times)^2$ (see [2, 22.10]), so $\Omega_n(F) = \mathrm{SO}_n(F)$ if $F = \bar{F}$. Note that if $F = \bar{F}$, then the algebraic group $\mathrm{O}_n(F)$ is disconnected and $\Omega_n(F)$ is the connected component $\mathrm{O}_n(F)^\circ$ containing the identity.

(v) *Similarities.* If $g \in \mathrm{Sim}(V, \kappa)$, then there exists $\tau(g) \in F^\times$ such that for all $u, v \in V$ we have $(ug, vg) = \tau(g)(u, v)$ (or $Q(vg) = \tau(g)Q(v)$). We refer to $\tau\colon \mathrm{Sim}(V, \kappa) \to F^\times$ as the *similarity map*.

(vi) *Warning.* Although we use notation such as $\mathrm{GO}_{2m}^-(q)$, the elements of these groups are linear maps on a fixed vector space V which preserve a fixed quadratic or bilinear form; the elements are not matrices. Indeed, we will use a number of different bases to specify elements in these groups.

(vii) *Notation.* Our notation in Fig. 2.1 is consistent with [2, 11, 20, 37, 50], sources to which we often refer (in general we try to always be consistent with [37]). However, this notation is not universal (in particular, when $F = \bar{F}$, the notation $\Omega_n(F)$ for the simple algebraic group $\mathrm{O}_n(F)^\circ$ is nonstandard).

By a *finite simple classical group* we mean one of the groups in Fig. 2.2. These groups are simple and each excluded group is either not simple or is isomorphic to a simple group that is included [50, Theorem 2.1.3 and Proposition 2.9.1].

We conclude this section by coining a useful piece of notation. If p is odd, then $\det(g) = \pm\tau(g)^m$ for all $g \in \mathrm{GO}_{2m}^\varepsilon(F)$ (see [50, Lemma 2.8.4]) and we define

$$\mathrm{DO}_{2m}^\varepsilon(F) = \{g \in \mathrm{GO}_{2m}^\varepsilon(F) \mid \det(g) = \tau(g)^m\}. \tag{2.3}$$

Informally, $\mathrm{DO}_n^\varepsilon(F)$ is to $\mathrm{GO}_n^\varepsilon(F)$ as $\mathrm{SO}_n^\varepsilon(F)$ is to $\mathrm{O}_n^\varepsilon(F)$; indeed, we have that $\mathrm{DO}_n^\varepsilon(F) \cap \mathrm{O}_n^\varepsilon(F) = \mathrm{SO}_n^\varepsilon(F)$. If $p = 2$, then we simply define

$$\mathrm{DO}_{2m}^\varepsilon(F) = \Omega_{2m}^\varepsilon(F). \tag{2.4}$$

Fig. 2.2 Finite simple classical groups

	$\mathrm{PSL}_n(q)$	$\mathrm{PSU}_n(q)$	$\mathrm{PSp}_n(q)$	$\mathrm{P}\Omega_n^\varepsilon(q)$
lower bound on n	2	3	4	7
excluded (n, q)	$(2, 2), (2, 3)$	$(3, 2)$	$(4, 2)$	

2.3 Actions of Classical Groups

In this section, $V = F^n$ where $n \geqslant 1$ and F is a field. We begin by recording general results, which are surely well known but are hard to find direct references for.

Let $\mathcal{D}$ be a direct sum decomposition $V = V_1 \oplus \cdots \oplus V_k$ or a tensor product decomposition $V = V_1 \otimes \cdots \otimes V_k$, where $\dim V_i > 1$ in the latter case. For $G \leqslant \mathrm{GL}(V)$, the centraliser $G_{(\mathcal{D})}$ and stabiliser $G_{\mathcal{D}}$ of $\mathcal{D}$ are the subgroups that stabilise the factors $V_1, \ldots, V_k$ pointwise and setwise, respectively. If an element $g \in \mathrm{GL}(V)$ centralises the decomposition $\mathcal{D}$ and acts as g_i on V_i, then we write g as $g_1 \oplus \cdots \oplus g_k$ or $g_1 \otimes \cdots \otimes g_k$, according to the type of decomposition.

The following is entirely analogous to Goursat's Lemma from group theory (see [51, p.75] for example).

Lemma 2.3.1 *Let $G \leqslant \mathrm{GL}(V)$ centralise $V = V_1 \oplus V_2$. Let U be an FG-submodule of V. Then there exist FG-submodules $W_1 \leqslant U_1 \leqslant V_1$ and $W_2 \leqslant U_2 \leqslant V_2$ and an FG-isomorphism $\varphi \colon U_1/W_1 \to U_2/W_2$ such that*

$$U = \{(u_1, u_2) \in U_1 \oplus U_2 \mid \varphi(W_1 + u_1) = W_2 + u_2\}.$$

Corollary 2.3.2 *Let $G \leqslant \mathrm{GL}(V)$ centralise $V = V_1 \oplus V_2$. Assume that there are no nonzero FG-isomorphisms between FG-subquotients of V_1 and V_2. Let U be an FG-submodule of V. Then there exist FG-submodules $U_1 \leqslant V_1$ and $U_2 \leqslant V_2$ such that $U = U_1 \oplus U_2$.*

The following lemma, which is proved directly in [50, Lemma 2.10.11], is an immediate consequence of Corollary 2.3.2.

Lemma 2.3.3 *Let $G \leqslant \mathrm{GL}(V)$ centralise $V = V_1 \oplus \cdots \oplus V_k$. If $V_1, \ldots, V_k$ are pairwise nonisomorphic irreducible FG-modules, then they are the only irreducible FG-submodules of V.*

We use the following straightforward lemma to compute centralisers of elements in classical groups.

Lemma 2.3.4 *Let $g = g_1 \oplus \cdots \oplus g_k \in \mathrm{GL}(V)$ centralise $V = V_1 \oplus \cdots \oplus V_k$. If there are no nonzero $F\langle g\rangle$-homomorphisms between V_i and V_j when $i \neq j$, then*

$$C_{\mathrm{GL}(V)}(g) = C_{\mathrm{GL}(V_1)}(g_1) \times \cdots \times C_{\mathrm{GL}(V_k)}(g_k).$$

For the remainder of this section it will be convenient to fix a basis for V and consider the elements of $\mathrm{GL}_n(F)$ as matrices with respect to this basis. For $g \in \mathrm{GL}_n(F)$, if V is an irreducible $F\langle g\rangle$-module, then we say that g is *irreducible*.

Lemma 2.3.5 *Let $g \in \mathrm{GL}_n(F)$. Then g is irreducible if and only if the characteristic polynomial of g is irreducible over F.*

Proof Let χ be the characteristic polynomial of g. First assume that g is reducible. Then g is similar to the block lower triangular matrix

$$\begin{pmatrix} g_1 & 0 \\ h & g_2 \end{pmatrix}$$

where g_1 is a $k \times k$ matrix for some $0 < k < n$. Therefore, the characteristic polynomial ϕ of g_1 is a proper nonconstant divisor of χ, so χ is reducible.

For the converse, assume that g is irreducible. From the rational canonical form of g, it is evident that the irreducibility of g implies that χ is the minimal polynomial of g. We wish to prove that χ is irreducible, so write $\chi = \phi\psi$, where ϕ and ψ are monic. Since $\chi(g) = 0$, without loss of generality, $\phi(g)$ is not invertible. Now let U be the kernel of $\phi(g)$, noting that $U \neq 0$. Let $u \in U$ and note that $(ug)\phi(g) = (u\phi(g))g = 0g = 0$, so U is a submodule of V. However, V is irreducible, so $U = V$ and, consequently, $\phi(g) = 0$. Since χ is the minimal polynomial of g, we deduce that $\chi = \phi$. Therefore, χ is irreducible. This completes the proof. □

Lemma 2.3.6 *Let $g, h \in \mathrm{GL}_n(F)$ be irreducible. Then g and h are similar if and only if they have the same characteristic polynomial.*

Proof If g and h are similar, then g and h evidently have the same characteristic polynomial. Now assume χ is the characteristic polynomial of both g and h. By Lemma 2.3.5, χ is irreducible. Therefore, the rational canonical form of both g and h is the companion matrix of χ, so g and h are similar. □

We say that an element $g \in \mathrm{GL}_n(F)$ is semisimple if g is similar to a block diagonal matrix $g_1 \oplus \cdots \oplus g_k$ where each g_i is irreducible.

Lemma 2.3.7 *Let $g, h \in \mathrm{GL}_n(F)$ be semisimple. Then g and h are similar if and only if they have the same characteristic polynomial.*

Proof Assume that χ is the characteristic polynomial of both g and h. Since g and h are semisimple, they are similar to block diagonal matrices $g_1^{a_1} \oplus \cdots \oplus g_k^{a_k}$ and $h_1^{b_1} \oplus \cdots \oplus h_l^{b_l}$, where $g_1, \ldots, g_k$ and $h_1, \ldots, h_l$ are pairwise non-similar irreducible matrices. For each i, let ϕ_i and ψ_i be the characteristic polynomials of g_i and h_i, respectively. By Lemma 2.3.5, the polynomials ϕ_i and ψ_i are irreducible since the matrices g_i and h_i are irreducible. Now $\phi_1^{a_1} \cdots \phi_k^{a_k} = \chi = \psi_1^{b_1} \cdots \psi_l^{b_l}$. By the irreducibility of each ϕ_i and ψ_i, we conclude $k = l$ and we may assume that for each i we have $\phi_i = \psi_i$ and $a_i = b_i$. For each i, by Lemma 2.3.6, g_i and h_i are similar since g_i and h_i are irreducible and have equal characteristic polynomials. Therefore, g and h are similar, as required. □

2.4 Standard Bases

Let F be a finite or algebraically closed field of characteristic $p > 0$ and $V = F^n$. We now fix standard bases for V for each classical form, following [50, Chapter 2].

First assume that $\kappa = (\cdot,\cdot)$ is symplectic. Fix $\mathcal{B} = (e_1, f_1, \ldots, e_m, f_m)$ such that

$$(e_i, e_j) = (f_i, f_j) = 0, \quad (e_i, f_j) = \delta_{ij}. \tag{2.5}$$

Next assume that $\kappa = Q$ is a nondegenerate quadratic form with associated bilinear form $(\cdot,\cdot)$. If $n = 2m+1$ is odd, then fix $\mathcal{B} = (e_1, f_1, \ldots, e_m, f_m, x)$ such that

$$Q(e_i) = Q(f_i) = 0, \quad Q(x) = 1, \quad (e_i, f_j) = \delta_{ij}, \quad (e_i, x) = (f_i, x) = 0. \tag{2.6}$$

Now assume that $n = 2m$ is even. If $\mathrm{sgn}(Q) = +$, then fix $\mathcal{B}^+ = (e_1, f_1, \ldots, e_m, f_m)$ such that

$$Q^+(e_i) = Q^+(f_i) = 0, \quad (e_i, f_j) = \delta_{ij}. \tag{2.7}$$

If $F = \mathbb{F}_q$ and $\mathrm{sgn}(Q) = -$, then, deviating from [50] and following [37], fix $\mathcal{B}^- = (e_1, f_1, \ldots, e_{m-1}, f_{m-1}, u_m, v_m)$ such that

$$\begin{aligned} &Q^-(e_i) = Q^-(f_i) = (e_i, u_m) = (f_i, u_m) = (e_i, v_m) = (f_i, v_m) = 0, \\ &(e_i, f_j) = \delta_{ij}, \quad Q^-(u_m) = Q^-(v_m) = \xi^{q+1}, \quad (u_m, v_m) = \xi^2 + \xi^{-2} \end{aligned} \tag{2.8}$$

where $\xi \in \mathbb{F}_{q^2} \setminus \mathbb{F}_q$ satisfies $|\xi| = q+1$ if $q \neq 3$ and $|\xi| = 8$ if $q = 3$. (Note that when $q = 3$ our definition of the minus-type standard basis corrects that in [37, Section 2.7], where the basis given there is not linearly independent.)

Finally assume that $F = \mathbb{F}_{q^2}$ and $\kappa = (\cdot,\cdot)$ is unitary. We fix two bases. First

$$\mathcal{B}_0 = (u_1, \ldots, u_n) \tag{2.9}$$

where $(u_i, u_j) = \delta_{ij}$. For the second basis, let $m = \lfloor \frac{n}{2} \rfloor$ and fix $e_i = u_{2i-1} + \zeta u_{2i}$ and $f_i = \zeta u_{2i-1} + u_{2i}$ where $\zeta \in \mathbb{F}_{q^2}^{\times}$ satisfies $\zeta^2 - \zeta - 1 = 0$. Note that

$$(e_i, e_j) = (f_i, f_j) = 0, \quad (e_i, f_j) = \delta_{ij}, \quad (e_i, u_n) = (f_i, u_n) = 0 \text{ if } n \text{ is odd.}$$

Let $\alpha, \beta \in \mathbb{F}_{q^2}^{\times}$ satisfy $\alpha^{q-1} = -1$ and $\beta^{q+1} = (-1)^m$ (choose $\alpha = \beta = 1$ if $p = 2$), and write

$$\mathcal{B} = \begin{cases} (\alpha e_1, -\alpha e_2, \ldots, (-1)^{m+1}\alpha e_m, f_m, \ldots, f_1) & \text{if } n \text{ is even} \\ (e_1, -e_2, \ldots, (-1)^{m+1} e_m, \beta u_n, f_m, \ldots, f_1) & \text{if } n \text{ is odd.} \end{cases} \tag{2.10}$$

2.5 Classical Algebraic Groups

The finite simple groups of Lie type arise from fixed points of algebraic groups under Steinberg endomorphisms (see [37, Chapters 1 and 2]), and this perspective allows us to exploit Shintani descent, which is described in Chap. 3.

Fix a prime p. By an *algebraic group* we always mean a linear algebraic group over $\overline{\mathbb{F}}_p$. For an indecomposable root system Φ, there exist simple algebraic groups Φ^{sc} and Φ^{ad} of simply connected and adjoint types, respectively, such that $Z(\Phi^{\mathrm{sc}})$ is finite, $Z(\Phi^{\mathrm{ad}}) = 1$ and $\Phi^{\mathrm{ad}} = \Phi^{\mathrm{sc}}/Z(\Phi^{\mathrm{sc}})$. Moreover, if X is a simple algebraic group with root system Φ, then there exist isogenies $\Phi^{\mathrm{sc}} \to X \to \Phi^{\mathrm{ad}}$ (see [37, Theorem 1.10.4]). For ease of notation, we refer to Φ^{ad} as Φ.

The classical algebraic groups are given in Fig. 2.3 (see [37, Theorem 1.10.7]). We adopt the notation introduced in Sect. 2.2 but omit reference to the field $\overline{\mathbb{F}}_p$, so $\mathrm{SO}_n = \mathrm{O}_n \cap \mathrm{SL}_n$ and $\Omega_n = \mathrm{O}_n^\circ$. Note that $\mathsf{D}_m^{\mathrm{ad}} = \mathrm{PDO}_{2m}$ in all cases. (The groups of type B_m with $p = 2$ were not introduced in Sect. 2.2. We include them here for completeness, but they will not feature later as they are not isometry groups of nondegenerate forms and they are isogenous to the groups of type C_m with $p = 2$.)

By a *Steinberg endomorphism* of an algebraic group X, we mean a bijective morphism $\sigma\colon X \to X$ whose fixed point subgroup $X_\sigma = \{x \in X \mid x^\sigma = x\}$ is finite.

Let X be a simple algebraic group of adjoint type and let σ be a Steinberg endomorphism of X. Then $T = O^{p'}(X_\sigma)$ is typically a finite simple group (see [37, Theorem 2.2.7(a)]) and the groups obtained in this way are the *finite simple groups of Lie type*. In this notation, the *innerdiagonal group* of T is $\mathrm{Inndiag}(T) = X_\sigma$.

Definition 2.5.1 Let $\mathcal{B}$ be a basis for $\overline{\mathbb{F}}_p^n$ and write the elements of $\mathrm{GL}_n(\overline{\mathbb{F}}_p)$ as matrices with respect to $\mathcal{B}$.

Φ	p	Φ^{sc}	Φ^{ad}	$\lvert Z(\Phi^{\mathrm{sc}})\rvert$	σ	$(\Phi^{\mathrm{ad}})_\sigma$
A_m $(m \geqslant 1)$		SL_{m+1}	PSL_{m+1}	$(m+1)_{p'}$	φ^f	$\mathrm{PGL}_{m+1}(q)$
					$\gamma\varphi^f$	$\mathrm{PGU}_{m+1}(q)$
B_m $(m \geqslant 2)$	2	SO_{2m+1}	SO_{2m+1}	1		
	odd	Spin_{2m+1}	SO_{2m+1}	2	φ^f	$\mathrm{PSO}_{2m+1}(q)$
C_m $(m \geqslant 2)$		Sp_{2m}	PSp_{2m}	$(p-1, 2)$	φ^f	$\mathrm{PGSp}_{2m}(q)$
D_m $(m \geqslant 4)$	2	Ω_{2m}	Ω_{2m}	1	φ^f	$\Omega^+_{2m}(q)$
					$r\varphi^f$	$\Omega^-_{2m}(q)$
	odd	Spin_{2m}	PSO_{2m}	4	φ^f	$\mathrm{PDO}^+_{2m}(q)$
					$r\varphi^f$	$\mathrm{PDO}^-_{2m}(q)$

Note: φ is $\varphi_{\mathcal{B}}$ or $\varphi_{\mathcal{B}^+}$, as appropriate, where $\mathcal{B}$ and $\mathcal{B}^+$ are defined in Sect. 2.4

Fig. 2.3 Simple classical algebraic groups

(i) The *standard Frobenius endomorphism* of $\mathrm{GL}_n(\overline{\mathbb{F}}_p)$ with respect to $\mathcal{B}$ is $\varphi_{\mathcal{B}}\colon (x_{ij}) \mapsto (x_{ij}^p)$.
(ii) The *standard graph automorphism* of $\mathrm{GL}_n(\overline{\mathbb{F}}_p)$ with respect to $\mathcal{B}$ is the map $\gamma_{\mathcal{B}}\colon x \mapsto (x^{-\mathsf{T}})^J$, where J is the antidiagonal matrix with entries $1, -1, 1, -1, \ldots, (-1)^{n+1}$ (from top-right to bottom-left).
(iii) Let $n = 2m$ and $\mathcal{B} = \mathcal{B}^+$ (from 2.7). The *standard reflection* $r \in \mathrm{O}_n^+(\overline{\mathbb{F}}_p)$ is

$$r = I_{n-2} \perp \begin{pmatrix} 0 & 1 \\ 1 & 0 \end{pmatrix}$$

that centralises the decomposition $\langle e_1, \ldots, f_{m-1}\rangle \perp \langle e_m, f_m\rangle$. Identify r with the *standard graph automorphism* of $\mathrm{GO}_n(\overline{\mathbb{F}}_p)$ that it induces by conjugation.

Observe that each of the graph automorphisms defined in parts (ii) and (iii) of Definition 2.5.1 are involutions and they commute with the standard Frobenius endomorphism defined in part (i).

Remark 2.5.2 Let us allow two notational conveniences.

(i) If the basis $\mathcal{B}$ is understood, then we write $\varphi = \varphi_{\mathcal{B}}$.
(ii) If θ is an endomorphism defined in Definition 2.5.1, then we will identify θ with the map induced on θ-stable subgroups of $\mathrm{GL}_n(\overline{\mathbb{F}}_p)$ and quotients of such subgroups by θ-stable normal subgroups.

Fix $f \geqslant 1$ and write $q = p^f$. If X is a simple classical algebraic group of adjoint type and σ is a Steinberg endomorphism of X that appears in the sixth column of Fig. 2.3, then the isomorphism type of X_σ is given in the seventh column of Fig. 2.3. This is essentially proved in [37, Section 2.7], but we will provide some of the details of the proof, since it will be important later that we understand the group X_σ exactly, not just up to isomorphism.

Lemma 2.5.3 *Let X be a simple classical algebraic group of adjoint type and let σ be a Steinberg endomorphism of X that appears in the sixth column of Fig. 2.3. Assume that $(X, \sigma) \neq (\mathsf{D}_m, r\varphi^f)$. Then X_σ is the group in the seventh column.*

Proof Assume (X, σ) is $(\mathsf{A}_m, \gamma\varphi^f)$ or $(\mathsf{D}_m, \varphi^f)$; the other cases are similar.

First let $(X, \sigma) = (\mathsf{A}_m, \gamma\varphi^f)$. Let $X = \mathrm{PSL}_n$ and write the elements of X with respect to a fixed basis $\mathcal{B} = (v_1, \ldots, v_n)$ for $\overline{\mathbb{F}}_p^n$. Write $Z = Z(\mathrm{GL}_n(\overline{\mathbb{F}}_p)) \cong \overline{\mathbb{F}}_p^\times$. For each $\mu \in \overline{\mathbb{F}}_p$, there exists $\lambda \in \overline{\mathbb{F}}_p$ such that $\lambda^n = \mu$ and hence there exists $\lambda I_n \in Z$ such that $\det(\lambda I_n) = \mu$. Consequently,

$$X = \mathrm{PSL}_n(\overline{\mathbb{F}}_p) = (\mathrm{SL}_n(\overline{\mathbb{F}}_p)Z)/Z = \mathrm{GL}_n(\overline{\mathbb{F}}_p)/Z = \mathrm{PGL}_n(\overline{\mathbb{F}}_p).$$

If $x \in X_{\gamma\varphi^f}$, then $x \in X_{\varphi^{2f}} = \mathrm{PGL}_n(q^2)$. Moreover, for $x \in \mathrm{PGL}_n(q^2)$

$$x \in X_{\gamma\varphi^f} \iff xJx^{\mathsf{T}\varphi^f} = J \iff x\alpha J x^{\mathsf{T}\varphi^f} = \alpha J,$$

where J is the antidiagonal matrix from Definition 2.5.1(ii) and $\alpha \in \mathbb{F}_{q^2}^{\times}$ satisfies $\alpha^{q-1} = -1$. Observe that the Gram matrix of the nondegenerate unitary form with respect to the basis $\mathcal{B}$ in (2.10) is J if n is odd and αJ if n is even, so $X_{\gamma\varphi^f} = \mathrm{PGU}_n(q)$ in both cases.

Now let $(X, \sigma) = (\mathsf{D}_m, \varphi^f)$. First assume that $p = 2$. Since $X \leqslant \mathrm{O}_n(\overline{\mathbb{F}}_2)$, we have $X_{\varphi^f} \leqslant \mathrm{O}_n^+(q)$. Since $\Omega_n(\overline{\mathbb{F}}_2)$ does not contain any reflections, $X_{\varphi^f} \leqslant \Omega_n^+(q)$. However, $|\,\mathrm{O}_n(\overline{\mathbb{F}}_2) : \Omega_n(\overline{\mathbb{F}}_2)| = 2$, so $|\,\mathrm{O}_n^+(q) : X_{\varphi^f}| \leqslant 2$. Therefore, $X_{\varphi^f} = \Omega_n^+(q)$.

Now assume that p is odd. Write $Z = Z(\mathrm{GO}_n(\overline{\mathbb{F}}_p)) \cong \overline{\mathbb{F}}_p^{\times}$. Since $\det(\lambda I_n) = \lambda^n = \tau(\lambda I_n)^{n/2}$, we have $Z \leqslant \mathrm{DO}_n(\overline{\mathbb{F}}_p)$. Moreover, for each $\mu \in \overline{\mathbb{F}}_p$, there exists $\lambda \in \overline{\mathbb{F}}_p$ such that $\lambda^2 = \mu$ and hence there exists $\lambda I_n \in Z$ such that $\tau(\lambda I_n) = \mu$ and $\det(\lambda I_n) = \mu^{n/2}$. Consequently, $\mathrm{SO}_n(\overline{\mathbb{F}}_p)Z = \mathrm{DO}_n(\overline{\mathbb{F}}_p)$ and

$$X = \mathrm{PSO}_n(\overline{\mathbb{F}}_p) = (\mathrm{SO}_n(\overline{\mathbb{F}}_p)Z)/Z = \mathrm{DO}_n(\overline{\mathbb{F}}_p)/Z = \mathrm{PDO}_n(\overline{\mathbb{F}}_p), \tag{2.11}$$

whence $X_{\varphi^f} = \mathrm{PDO}_n^+(q)$. □

Next we handle the minus-type orthogonal groups.

Lemma 2.5.4 *Let $X = \mathsf{D}_m$ with $m \geqslant 4$ and let $\varphi = \varphi_{\mathcal{B}^+}$. Then there exists an inner automorphism Ψ of $\mathrm{GL}_{2m}(\overline{\mathbb{F}}_p)$ such that $X_{r\varphi^f}\Psi$ is $\mathrm{PDO}_{2m}^-(q)$.*

Proof Let $V = \overline{\mathbb{F}}_p^n$ be equipped with the quadratic form Q, with bilinear form $(\cdot,\cdot)$, defined in (2.7) with respect to the basis $\mathcal{B}^+ = (e_1, f_1, \dots, e_m, f_m)$, where $n = 2m$. Let Ψ be the automorphism of $\mathrm{GL}_n(\overline{\mathbb{F}}_p)$ induced by conjugation by the element $A = I_{n-2} \perp A'$ that centralises $\langle e_1, \dots, f_{m-1}\rangle \perp \langle e_m, f_m\rangle$, where

$$A' = \begin{pmatrix} \xi & \xi^{-1} \\ \xi^{-1} & \xi \end{pmatrix}$$

and where $\xi \in \mathbb{F}_{q^2} \setminus \mathbb{F}_q$ satisfies $|\xi| = q+1$ if $q \neq 3$ and $|\xi| = 8$ if $q = 3$.

Write $u_m = e_m A$ and $v_m = f_m A$. It is straightforward to check that $Q(u_m) = Q(v_m) = \xi^{q+1}$ and $(u_m, v_m) = \xi^2 + \xi^{-2}$, so, without loss of generality, we may assume that $\mathcal{B}^+ A$ is the basis $\mathcal{B}^-$ defined in (2.8).

Let $\sigma_\varepsilon = (\varphi_{\mathcal{B}^\varepsilon})^f$. A straightforward calculation yields $AA^{-(q)} = r$ where $A = (a_{ij})$ and $A^{(q)} = (a_{ij}^q)$. Consequently, $X_{r\sigma_+}\Psi = A^{-1}X_{r\sigma_+}A = X_{\sigma_-}$ for any subgroup $X \leqslant \mathrm{GL}_n(\overline{\mathbb{F}}_p)$. Let V^ε be the $\mathbb{F}_q$-span of $\mathcal{B}^\varepsilon$. Then (V^ε, Q) is the ε-type formed space from (2.7) or (2.8). Therefore, if $X = \mathrm{SO}_n(\overline{\mathbb{F}}_p)$, then $X_{\sigma_+} = \mathrm{SO}_n^+(q)$ and $X_{r\sigma_+}\Psi = X_{\sigma_-} = \mathrm{SO}_n^-(q)$.

We are ready to prove the main claims of the lemma.

First assume that $p = 2$ and $X = \Omega_n(\overline{\mathbb{F}}_2)$. We know that $Y_{r\varphi^f}\Psi = \mathrm{O}_n^-(q)$, where $Y = \mathrm{O}_n(\overline{\mathbb{F}}_2)$. Since Ψ^{-1} maps the reflections in $\mathrm{O}_n^-(q)$ to reflections in $Y_{r\varphi^f}$ and X contains no reflections, we conclude that $X_{r\varphi^f}\Psi = \Omega_n^-(q)$.

Now assume that p is odd and $X = \mathrm{PSO}_n(\overline{\mathbb{F}}_p)$. We recorded in (2.11) that $X = \mathrm{PSO}_n(\overline{\mathbb{F}}_p) = \mathrm{PDO}_n(\overline{\mathbb{F}}_p)$. The above discussion now implies that $X_{r\varphi^f\Psi} = \mathrm{PDO}_n^-(q)$. This completes the proof. □

We now see the significance of the notation $\mathrm{DO}_n^{\pm}(q)$ introduced in Sect. 2.2. Namely, in light of Lemmas 2.5.3 and 2.5.4, with a slight abuse of notation for minus-type groups, for even $n \geqslant 8$,

$$\mathrm{Inndiag}(\mathrm{P}\Omega_n^{\pm}(q)) = \mathrm{PDO}_n^{\pm}(q). \tag{2.12}$$

2.6 Maximal Subgroups of Classical Groups

An understanding of the subgroup structure of almost simple classical groups will be essential in Chaps. 5 and 6. Let G be an almost simple classical group and let $V = \mathbb{F}_{q^d}^n$ be the natural module for $\mathrm{soc}(G)$, where $q = p^f$ and $d \in \{1, 2\}$ (here $d = 2$ if and only if $\mathrm{soc}(G) = \mathrm{PSU}_n(q)$). Theorem 2.6.1 was proved by Aschbacher [1], except for the special case when $\mathrm{soc}(G) = \mathrm{P}\Omega_8^+(q)$ and G contains a triality automorphism, which was proved by Kleidman [47].

Theorem 2.6.1 *Let G be an almost simple classical group and let H be a maximal subgroup of G not containing* $\mathrm{soc}(G)$. *Then H belongs to one of the subgroup collections $\mathcal{C}_1, \dots, \mathcal{C}_8, \mathcal{S}, \mathcal{N}$.*

Regarding Theorem 2.6.1, notice that the subgroups of G that contain $\mathrm{soc}(G)$ correspond to subgroups of $G/\mathrm{soc}(G) \leqslant \mathrm{Out}(\mathrm{soc}(G))$, which is a well-known soluble group. This explains our focus on maximal subgroups not containing $\mathrm{soc}(G)$.

The collections $\mathcal{C}_1, \dots, \mathcal{C}_8$ contain the *geometric subgroups*, and each such collection corresponds to a different geometric structure on the natural module V. We adopt the definition of each $\mathcal{C}_i$ given in [50, Section 4.i], which differs slightly from Aschbacher's original definition. These eight collections are summarised in Fig. 2.4. Each $\mathcal{C}_i$ collection is a union of *types* of geometric subgroup. The type of a subgroup is a rough indication of both its group theoretic structure and the geometric structure it stabilises; this notion is formally introduced in [50, p.58]. The main theorem in [50, Chapter 3] establishes the structure, conjugacy and, when $n \geqslant 13$, maximality of each geometric subgroup of each almost simple classical group. Complete information on the maximal subgroups of almost simple classical groups with $n \leqslant 12$ is given in [8].

If $H \leqslant G$ is contained in the collection $\mathcal{S}$, then H is almost simple with socle H_0 and the embedding $H_0 \leqslant G$ is afforded by an absolutely irreducible representation $\widehat{H}_0 \to \mathrm{GL}(V)$ for some quasisimple extension $\widehat{H}_0$ of H_0. If $\mathrm{soc}(G)$ is $\mathrm{Sp}_4(2^f)$ or $\mathrm{P}\Omega_8^+(q)$, then additional subgroups arise in a collection $\mathcal{N}$, described in [20, Table 5.9.1]; a feature of the subgroups $H \in \mathcal{N}$ is that they are *novelties*, that is, $H \cap \mathrm{soc}(G)$ is not maximal in $\mathrm{soc}(G)$.

	structure stabilised	rough description in $\mathrm{GL}_n(q)$
$\mathcal{C}_1$	n.d. or totally singular subspace	maximal parabolic
$\mathcal{C}_2$	$V = \bigoplus_{i=1}^k V_i$ where $\dim V_i = a$	$\mathrm{GL}_a(q) \wr S_k$ with $n = ak$
$\mathcal{C}_3$	prime degree field extension of $\mathbb{F}_q$	$\mathrm{GL}_a(q^k).k$ with $n = ak$ for prime k
$\mathcal{C}_4$	tensor product $V = V_1 \otimes V_2$	$\mathrm{GL}_a(q) \circ \mathrm{GL}_b(q)$ with $n = ab$
$\mathcal{C}_5$	prime degree subfield of $\mathbb{F}_q$	$\mathrm{GL}_n(q_0)$ with $q = q_0^k$ for prime k
$\mathcal{C}_6$	symplectic-type r-group	$(C_{q-1} \circ r^{1+2a}).\mathrm{Sp}_{2a}(r)$ with $n = r^a$
$\mathcal{C}_7$	$V = \bigotimes_{i=1}^k V_i$ where $\dim V_i = a$	$(\mathrm{GL}_a(q) \circ \cdots \circ \mathrm{GL}_a(q)).S_k$ with $n = a^k$
$\mathcal{C}_8$	nondegenerate classical form	$\mathrm{GSp}_n(q)$, $\mathrm{GO}_n^{\varepsilon}(q)$, $\mathrm{GU}_n(q^{\frac{1}{2}})$

Fig. 2.4 Geometric subgroups

A key aspect of the proofs in Chaps. 5 and 6 is to determine which maximal subgroups of a given almost simple classical group G contain a carefully chosen element $s \in G$. While we cannot typically use the order of s to do this (see Remark 5.4.9), when we can, we use the main theorem of [40], which we now discuss.

Let $a, b \geqslant 2$ be positive integers. Then the *primitive part* of $a^b - 1$, written $\Phi_b^*(a)$, is the largest divisor of $a^b - 1$ that is coprime to $a^k - 1$ for all $k < b$, and a *primitive prime divisor* of $a^b - 1$ is any prime divisor of $\Phi_b^*(a)$. The following is due to Zsigmondy [63] (see also [20, Theorem 3.1.5]).

Theorem 2.6.2 *Let a and b be positive integers. Then $\Phi_b^*(a) > 1$ if*

$$a \geqslant 2 \text{ and } (a, b) \neq (2, 6) \text{ and } a + 1 \text{ is not a power of } 2 \text{ if } b = 2. \tag{2.13}$$

The main theorem of [40] describes the maximal subgroups of $\mathrm{GL}_n(q)$ that contain an element whose order is divisible by a primitive prime divisor of $q^k - 1$ for $k > \frac{n}{2}$. We will use the more focussed version given in [39, Theorem 2.2] (the exceptions in part (iv) are given in [39, Table 1]).

Theorem 2.6.3 *Let $n \geqslant 3$ and $k > \frac{n}{2}$. Let $g \in \mathrm{GL}(V)$ and assume that $|g| > 2k+1$ divides $\Phi_k^*(q)$. Let $H \leqslant \mathrm{GL}(V)$ be an irreducible subgroup containing g. Then one of the following holds*

(i) *H is a subfield subgroup*
(ii) *H is a field extension subgroup of degree dividing (n, k)*
(iii) *H contains $\mathrm{SL}(V)$, $\mathrm{SU}(V)$, $\mathrm{Sp}(V)$ or $\Omega(V)$*
(iv) *$n \leqslant 9$ and H is one of a small number of exceptions.*

Theorem 2.6.3 raises the question of when $\Phi_e^*(q) > 2e + 1$, which is answered by the following lemma (see [5, Lemma 6.1]).

Lemma 2.6.4 *Let q and e be positive integers. Then $\Phi_e^*(q) \equiv 1 \pmod{e}$. Moreover, if q is a prime power and $e \geqslant 3$, then*

$$\Phi_e^*(q) \begin{cases} = 1 & \text{if } (q, e) = (2, 6) \\ = e + 1 & \text{if } (q, e) \in \{(2, 4), (2, 10), (2, 12), (2, 18), (3, 4), (3, 6), (5, 6)\} \\ = 2e + 1 & \text{if } (q, e) \in \{(2, 3), (2, 8), (2, 20), (4, 3), (4, 6)\} \\ \geqslant 3e + 1 & \text{otherwise.} \end{cases}$$

2.7 Computational Methods

Based on the work of Breuer in [10], we implemented an algorithm in Magma [6] that takes as input a finite group G, an element $s \in G$ and nonnegative integers k and N, with the aim of determining whether s^G witnesses $u(G) \geqslant k$. An overview of this algorithm is given in [43, Section 2.3] and the relevant code is in Appendix A. The computations were carried out in Magma 2.24-4 on a 2.7 GHz machine with 128 GB RAM. The largest computation took 472 s and 417 MB of memory, and this was to prove that $u(\langle \Omega_8^+(4), \theta \rangle) \geqslant 2$ where θ is an involutory field automorphism.

Chapter 3
Shintani Descent

In this chapter, we describe Shintani descent, which is an important technique for understanding the conjugacy classes in almost simple groups. Shintani descent is crucial to this project and also useful more generally. In Sect. 3.1, we follow the account given in [21, Section 2.6], and Sect. 3.3 records some of the key applications of Shintani descent. With the goal of providing a self-contained reference for future use, we have decided to give our own treatment of these existing results and we take the opportunity to set these results in a general context.

However, for our application, the existing techniques of Shintani descent are not sufficient and we need to develop the theory in order to handle, for example, twisted groups of Lie type. In particular, Sect. 3.2 features three new technical lemmas on Shintani maps, and Sect. 3.4 introduces a new result that allows us to use Shintani descent in contexts that previously were not amenable to this approach.

3.1 Introduction

For this entire chapter, let X be a connected algebraic group over $\overline{\mathbb{F}}_p$ and let σ be a Steinberg endomorphism of X. The following is the Lang–Steinberg Theorem [60, Theorem 10.13].

Theorem 3.1.1 *The map $L\colon X \to X$ defined as $L(x) = xx^{-\sigma}$ is surjective.*

Corollary 3.1.2 *The map $L'\colon X \to X$ where $L'(x) = xx^{-\sigma^{-1}}$ is surjective.*

Proof Let $g \in X$. Theorem 3.1.1 implies that there exists $x \in X$ such that $g^{-\sigma} = xx^{-\sigma}$. Consequently, $g = xx^{-\sigma^{-1}}$ and L' is surjective. $\square$

Fix an integer $e > 1$. The subgroup X_{σ^e} is σ-stable, so σ restricts to an automorphism $\widetilde{\sigma} = \sigma|_{X_{\sigma^e}}$ of X_{σ^e}, and we can consider the finite semidirect product $X_{\sigma^e}{:}\langle\widetilde{\sigma}\rangle$, where $g^{\widetilde{\sigma}} = g^{\sigma}$ for all $g \in X_{\sigma^e}$, noting that $|\widetilde{\sigma}| = e$.

S. Harper, *The Spread of Almost Simple Classical Groups*, Lecture Notes in Mathematics 2286, https://doi.org/10.1007/978-3-030-74100-6_3

Definition 3.1.3 A *Shintani map* of (X, σ, e) is a map of conjugacy classes

$$F : \{(g\widetilde{\sigma})^{X_{\sigma^e}} \mid g \in X_{\sigma^e}\} \to \{x^{X_\sigma} \mid x \in X_\sigma\} \quad (g\widetilde{\sigma})^{X_{\sigma^e}} \mapsto (a^{-1}(g\widetilde{\sigma})^e a)^{X_\sigma}$$

where $a \in X$ satisfies $g = aa^{-\sigma^{-1}}$ (which exists by Corollary 3.1.2).

We will often abuse notation by using $F(g\widetilde{\sigma})$ to refer to a representative of the X_σ-class $F((g\widetilde{\sigma})^{X_{\sigma^e}})$.

The following theorem establishes the main properties of the Shintani map. It was first proved by Kawanaka in [45], building on earlier work of Shintani who introduced the key ideas in [58]. We follow the proof of [21, Lemma 2.13].

Theorem 3.1.4 (Shintani Descent) *Let F be a Shintani map of (X, σ, e).*

(i) *The map F is a well-defined bijection, independent of the choice of $a \in X$.*
(ii) *If $g \in X_{\sigma^e}$ then $C_{X_\sigma}(F(g\widetilde{\sigma})) = a^{-1}C_{X_{\sigma^e}}(g\widetilde{\sigma})a$.*

Proof Let $g \in X_{\sigma^e}$ and write $g = aa^{-\sigma^{-1}}$. First note that

$$\begin{aligned} a^{-1}(g\widetilde{\sigma})^e a &= a^{-1} g g^{\sigma^{-1}} \cdots g^{\sigma^{-(e-1)}} a \\ &= a^{-1}(aa^{-\sigma^{-1}})(a^{\sigma^{-1}}a^{-\sigma^{-2}}) \cdots (a^{\sigma^{-(e-1)}}a^{-\sigma^{-e}})a = a^{-\sigma^{-e}}a. \end{aligned}$$

Since $g = aa^{-\sigma^{-1}} \in X_{\sigma^e}$ we know that $aa^{-\sigma^{-1}} = (aa^{-\sigma^{-1}})^{\sigma^{-e}} = a^{\sigma^{-e}}a^{-\sigma^{-(e+1)}}$, whence $a^{-\sigma^{-e}}a = a^{-\sigma^{-(e+1)}}a^{\sigma^{-1}} = (a^{-\sigma^{-e}}a)^{\sigma^{-1}}$, so $a^{-\sigma^{-e}}a \in X_\sigma$.

Let $h\widetilde{\sigma}$ be X_{σ^e}-conjugate to $g\widetilde{\sigma}$. Fix $k \in X_{\sigma^e}$ such that $h\widetilde{\sigma} = k^{-1}(g\widetilde{\sigma})k$ and consequently $h = k^{-1}gk^{-\sigma^{-1}}$. Writing $g = aa^{-\sigma^{-1}}$, we obtain $h = (k^{-1}a)(k^{-1}a)^{-\sigma^{-1}}$, whence

$$(k^{-1}a)^{-1}(h\widetilde{\sigma})^e(k^{-1}a) = a^{-1}k(h\widetilde{\sigma})^e k^{-1}a = a^{-1}(g\widetilde{\sigma})^e a.$$

Therefore, F does not depend on the choice of representative of the X_{σ^e}-class.

Write $g = aa^{-\sigma^{-1}} = bb^{-\sigma^{-1}}$. Then $a^{-1}b = a^{-\sigma^{-1}}b^{\sigma^{-1}} = (a^{-1}b)^{\sigma^{-1}}$, so $a^{-1}b \in X_\sigma$ and

$$b^{-1}(g\widetilde{\sigma})^e b = (a^{-1}b)^{-1}(a^{-1}(g\widetilde{\sigma})^e a)(a^{-1}b),$$

so F is independent of the choice of a. Therefore, F is a well-defined function.

To see that F is surjective, let $x \in X_\sigma$ and write $x^{-1} = bb^{-\sigma^{-e}}$. Therefore, writing $a = b^{-1}$, we have $x = a^{-\sigma^{-e}}a$. As we argued in the first paragraph, $a^{-1}(aa^{-\sigma^{-1}}\widetilde{\sigma})^e a = x$ and $aa^{-\sigma^{-1}} \in X_{\sigma^e}$ since $a^{-\sigma^{-e}}a \in X_\sigma$. We will complete the proof that F is bijective after proving (ii).

Turning to (ii), let $z \in C_{X_{\sigma^e}}(g\widetilde{\sigma})$. Then $a^{-1}za$ centralises $a^{-1}(g\widetilde{\sigma})^e a$. The fact that $z \in C_{X_{\sigma^e}}(g\widetilde{\sigma})$ implies that $zg\widetilde{\sigma} = g\widetilde{\sigma}z$, so $z^{\sigma^{-1}} = g^{-1}zg$. Therefore,

$$(a^{-1}za)^{\sigma^{-1}} = a^{-\sigma^{-1}}g^{-1}zga^{\sigma^{-1}} = a^{-1}gg^{-1}zgg^{-1}a = a^{-1}za.$$

Therefore, $a^{-1}za \in X_\sigma$, so $a^{-1}za \in C_{X_\sigma}(a^{-1}(g\widetilde{\sigma})^e a) = C_{X_\sigma}(F(g\widetilde{\sigma}))$. This proves that $a^{-1}C_{X_{\sigma^e}}(g\widetilde{\sigma})a \subseteq C_{X_\sigma}(F(g\widetilde{\sigma}))$. For the reverse inclusion, let $w \in C_{X_\sigma}(F(g\widetilde{\sigma}))$. Then

$$awa^{-1} = (g\widetilde{\sigma})^{-e}(awa^{-1})(g\widetilde{\sigma})^e = (aa^{-\sigma^{-1}}\sigma)^{-e}(awa^{-1})(aa^{-\sigma^{-1}}\sigma)^e = (awa^{-1})^{\sigma^{-e}}$$

which implies that $awa^{-1} \in X_{\sigma^e}$. Moreover,

$$(g\widetilde{\sigma})^{-1}(awa^{-1})(g\widetilde{\sigma}) = (\sigma^{-1}a^{\sigma^{-1}}a^{-1})awa^{-1}(aa^{-\sigma^{-1}}\sigma) = aw^{\sigma^{-1}}a^{-1} = awa^{-1},$$

so $awa^{-1} \in C_{X_{\sigma^e}}(g\widetilde{\sigma})$. This implies that $a^{-1}C_{X_{\sigma^e}}(g\widetilde{\sigma})a = C_{X_\sigma}(F(g\widetilde{\sigma}))$, as claimed.

We may now prove that F is bijective. Let $\{c_1, \dots, c_t\}$ be representatives of the X_σ-classes in X_σ. Then there exist X_{σ^e}-classes $C_1, \dots, C_t$ in $X_{\sigma^e}\widetilde{\sigma}$ such that $F(C_i) = c_i$ for each i, by the surjectivity of F. By (ii), $|C_i| = |c_i||X_{\sigma^e} : X_\sigma|$. This implies that

$$\sum_{i=1}^{t} |C_i| = |X_{\sigma^e} : X_\sigma| \sum_{i=1}^{t} |c_i| = |X_{\sigma^e}| = |X_{\sigma^e}\widetilde{\sigma}|,$$

so $\{C_1, \dots, C_t\}$ is the set of G-classes in $X_{\sigma^e}\widetilde{\sigma}$, which proves F is bijective. □

Theorem 3.1.4(i) explains why we refer to F as *the* Shintani map of (X, σ, e). The following concrete example highlights how we apply Shintani descent.

Example 3.1.5 Let $e \geqslant 2$, let $m \geqslant 4$ and let $q = 2^e$. Write $X = \Omega_{2m}(\overline{\mathbb{F}}_2)$. Let $\varphi = \varphi_{\mathcal{B}^+}$ be the standard Frobenius endomorphism $(a_{ij}) \mapsto (a_{ij}^2)$ of X.

Let F be the Shintani map of (X, φ, e). Note that $X_\varphi = \Omega_{2m}^+(2)$ and that $X_{\varphi^e} = \Omega_{2m}^+(q)$. Now

$$F\colon \{(g\varphi)^{\Omega_{2m}^+(q)} \mid g \in \Omega_{2m}^+(q)\} \to \{x^{\Omega_{2m}^+(2)} \mid x \in \Omega_{2m}^+(2)\}.$$

Therefore, we can specify a conjugacy class in the coset $\Omega_{2m}^+(q)\varphi$ of the almost simple group $\langle \Omega_{2m}^+(q), \varphi \rangle$ as the preimage under F of a conjugacy class in $\Omega_{2m}^+(2)$.

Recall the element r from Definition 2.5.1(iii). Let E be the Shintani map of $(X, r\varphi, e)$. Then $X_{r\varphi} \cong \Omega_{2m}^-(2)$ and $X_{(r\varphi)^e} \cong \Omega_{2m}^\varepsilon(q)$ where $\varepsilon = (-)^e$. Therefore, the map

$$E\colon \{(gr\varphi)^{\Omega_{2m}^\varepsilon(q)} \mid g \in \Omega_{2m}^\varepsilon(q)\} \to \{x^{\Omega_{2m}^-(2)} \mid x \in \Omega_{2m}^-(2)\}$$

allows us, for example, to specify elements in the coset $\Omega_{2m}^+(q)r\varphi$ of $\langle \Omega_{2m}^+(q), r\varphi \rangle$ when e is even. However, this setup does not shed light on this coset when e is odd. This is problematic as we will need to select an element in this coset in order to study the uniform spread of $\langle \Omega^+_{2m}(q), r\varphi \rangle$, and this shows the limitations of the current Shintani descent techniques. In Example 3.4.2, we will see how to handle this case using our new methods.

3.2 Properties

In this section, we will establish three new properties of the Shintani map, which justify techniques that we repeatedly employ. Each of these properties relies on the fact that the Shintani map does not depend on the choice of element afforded by the Lang–Steinberg Theorem (see Theorem 3.1.4(i)).

Throughout, we assume that X is a connected algebraic group, σ is a Steinberg endomorphism of X and $e > 1$. Let F be the Shintani map of (X, σ, e) and let $\widetilde{\sigma} = \sigma|_{X_{\sigma^e}}$.

We begin with a preliminary observation. If Y is a closed σ-stable subgroup of X, then the restriction σ_Y of σ to Y is a Steinberg endomorphism. Similarly, if $\pi : X \to Y$ is an isogeny with a σ-stable kernel, then σ induces a Steinberg endomorphism σ_Y on Y such that $\sigma_Y \circ \pi = \pi \circ \sigma$. For ease of notation, in both cases we write σ for σ_Y.

The first property concerns subgroups (an application is Proposition 5.4.11).

Lemma 3.2.1 *Let Y be a closed connected σ-stable subgroup of X and let E be the Shintani map of (Y, σ, e).*

(i) *For all $g \in Y_{\sigma^e}$, any representative of $E((g\widetilde{\sigma})^{Y_{\sigma^e}})$ represents $F((g\widetilde{\sigma})^{X_{\sigma^e}})$.*
(ii) *For all $x \in Y_\sigma$, any representative of $E^{-1}(x^{Y_\sigma})$ represents $F^{-1}(x^{X_\sigma})$.*

Proof We prove only (i) since (ii) is very similar. Let $g \in Y_{\sigma^e}$ and let x be a representative of $E((g\widetilde{\sigma})^{Y_{\sigma^e}})$. Then $x = a^{-1}(g\widetilde{\sigma})^e a$ for an element $a \in Y$ such that $aa^{-\sigma^{-1}} = g$. Since $Y \leqslant X$, the element $a^{-1}(g\widetilde{\sigma})^e a = x$ represents $F((g\widetilde{\sigma})^{X_{\sigma^e}})$. □

The second property concerns quotients.

Lemma 3.2.2 *Let $\pi : X \to Y$ be an isogeny with a σ-stable kernel and let E be the Shintani map of (Y, σ, e).*

(i) *For all $h \in \pi(X_{\sigma^e}) \leqslant Y_{\sigma^e}$, there exists $y \in \pi(X_\sigma) \leqslant Y_\sigma$ that represents the class $E(h\widetilde{\sigma})$.*
(ii) *For all $y \in \pi(X_\sigma) \leqslant Y_\sigma$, there exists $h \in \pi(X_{\sigma^e}) \leqslant Y_{\sigma^e}$ such that $h\widetilde{\sigma}$ represents the class $E^{-1}(y)$.*

Moreover, if $\langle \pi(X_{\sigma^e}), \widetilde{\sigma} \rangle \trianglelefteq \langle Y_{\sigma^e}, \widetilde{\sigma} \rangle$ and $\pi(X_\sigma) \trianglelefteq Y_\sigma$, then E restricts to a bijection

$$\{(h\widetilde{\sigma})^{Y_{\sigma^e}} \mid h \in \pi(X_{\sigma^e})\} \to \{y^{Y_\sigma} \mid y \in \pi(X_\sigma)\}.$$

Proof For (i), let $g \in X_{\sigma^e}$ and let x be a representative of $F(g\widetilde{\sigma})$. Then $x = a^{-1}(g\widetilde{\sigma})^e a$ for an element $a \in X$ such that $aa^{-\sigma^{-1}} = g$. Therefore, we have $\pi(x) = \pi(a)^{-1}(\pi(g)\widetilde{\sigma})^e\pi(a)$. Note that $\pi(x) \in \pi(X_\sigma) \leqslant Y_\sigma$. Moreover, $\pi(a) \in Y$ and $\pi(a)\pi(a)^{-\sigma^{-1}} = \pi(g)$, so $\pi(a)^{-1}(\pi(g)\widetilde{\sigma})^e\pi(a) = \pi(x)$ is a valid representative of $E(\pi(g)\widetilde{\sigma})$, as required. As with Lemma 3.2.1, (ii) is similar to (i).

If $\langle \pi(X_{\sigma^e}), \widetilde{\sigma} \rangle \trianglelefteqslant \langle Y_{\sigma^e}, \widetilde{\sigma} \rangle$ and $\pi(X_\sigma) \trianglelefteqslant Y_\sigma$, then for all $h \in \pi(X_{\sigma^e})$ and for all $y \in \pi(X_\sigma)$ we have $(h\widetilde{\sigma})^{Y_{\sigma^e}} \subseteq \pi(X_{\sigma^e})\widetilde{\sigma}$ and $y^{Y_\sigma} \subseteq \pi(X_\sigma)$, which implies, given (i) and (ii), that E restricts to the bijection in the statement. □

Corollary 3.2.3 *Let Y be a simple algebraic group of adjoint type, let σ be a Steinberg endomorphism of Y and let $e > 1$. Write $T = O^{p'}(Y_{\sigma^e})$ and assume that $\langle T, \widetilde{\sigma} \rangle \trianglelefteqslant \langle Y_{\sigma^e}, \widetilde{\sigma} \rangle$. Then the Shintani map of (Y, σ, e) restricts to a bijection*

$$\{(t\widetilde{\sigma})^{Y_{\sigma^e}} \mid t \in T\} \to \{y^{Y_\sigma} \mid y \in O^{p'}(Y_\sigma)\}.$$

Proof Let X be the simply connected version of Y, so $Y = X/Z(X)$, and let $\pi : X \to Y$ be the isogeny arising from taking the quotient by $Z(X)$. By [37, Theorem 2.1.2(e)], since X is simply connected, there is a unique Steinberg endomorphism σ_X of X such that $\pi \circ \sigma_X = \sigma \circ \pi$, so in particular, $\ker(\pi)$ is σ_X-stable. As usual, for ease of notation, we write $\sigma = \sigma_X$. We aim to apply Lemma 3.2.2, with the isogeny $\pi : X \to Y$ and the Shintani maps F and E of (X, σ, e) and (Y, σ, e), respectively. By [37, Proposition 2.5.9 and Theorem 2.2.6(c)],

$$\pi(X_\sigma) = X_\sigma/Z(X_\sigma) = O^{p'}(Y_\sigma) \trianglelefteqslant Y_\sigma \tag{3.1}$$

$$\pi(X_{\sigma^e}) = X_{\sigma^e}/Z(X_{\sigma^e}) = O^{p'}(Y_{\sigma^e}) = T, \tag{3.2}$$

and, by hypothesis, $\langle T, \widetilde{\sigma} \rangle \trianglelefteqslant \langle Y_{\sigma^e}, \widetilde{\sigma} \rangle$. Therefore, Lemma 3.2.2 implies that E restricts to the bijection

$$\{(t\widetilde{\sigma})^{Y_{\sigma^e}} \mid t \in T\} \to \{y^{Y_\sigma} \mid y \in O^{p'}(Y_\sigma)\}.$$

This completes the proof. □

The following example elucidates the utility of Corollary 3.2.3 and it provides an alternative proof of [43, Proposition 2.4] (see also Lemmas 5.4.3 and 5.4.4).

Example 3.2.4 Let $m \geqslant 2$, let p be an odd prime and let $q = q_0^e = p^f$, where $e \geqslant 2$ divides f. Write $Y = \mathrm{PSp}_{2m}(\overline{\mathbb{F}}_p)$ and let $\sigma = \varphi^{f/e}$ where $\varphi = \varphi_{\mathcal{B}^+}$ is the standard Frobenius endomorphism. The Shintani map E of (Y, σ, e) is

$$E : \{(g\widetilde{\sigma})^{\mathrm{PGSp}_{2m}(q)} \mid g \in \mathrm{PGSp}_{2m}(q)\} \to \{x^{\mathrm{PGSp}_{2m}(q_0)} \mid x \in \mathrm{PGSp}_{2m}(q_0)\}.$$

The map E allows us to identify a $\mathrm{PGSp}_{2m}(q)$-class $(g\widetilde{\sigma})^{\mathrm{PGSp}_{2m}(q)}$ in the coset $\mathrm{PGSp}_{2m}(q)\widetilde{\sigma}$ by specifying a conjugacy class $x^{\mathrm{PGSp}_{2m}(q_0)}$ of $\mathrm{PGSp}_{2m}(q_0)$, but we do not know which coset of $\mathrm{PSp}_{2m}(q)$ this class is contained in. However,

Corollary 3.2.3 implies that E restricts to a bijection

$$\{(t\widetilde{\sigma})^{\mathrm{PGSp}_{2m}(q)} \mid t \in \mathrm{PSp}_{2m}(q)\} \to \{y^{\mathrm{PGSp}_{2m}(q_0)} \mid y \in \mathrm{PSp}_{2m}(q_0)\},$$

which informs us that $g\widetilde{\sigma} \in \mathrm{PSp}_{2m}(q)\widetilde{\sigma}$ if and only if $x \in \mathrm{PSp}_{2m}(q_0)$.

We conclude with a property that relates Shintani maps to taking powers.

Lemma 3.2.5 *Let $x \in X_\sigma$ and assume that $F(g\widetilde{\sigma}) = x^{X_\sigma}$. Let $d > 1$.*

(i) *Let E_1 be the Shintani map of (X, σ^d, e). Then $E_1((g\widetilde{\sigma})^d) = (x^d)^{X_{\sigma^d}}$.*
(ii) *Assume d is a proper divisor of e and let E_2 be the Shintani map of $(X, \sigma^d, e/d)$. Then $E_2((g\widetilde{\sigma})^d) = x^{X_{\sigma^d}}$.*

Proof Assume that $g \in X_{\sigma^e}$ satisfies $F(g\widetilde{\sigma}) = x$. Fix an element $a \in X$ satisfying $a^{-1}(g\widetilde{\sigma})^e a = x$ and $aa^{-\sigma^{-1}} = g$. Write

$$h = gg^{\sigma^{-1}} \cdots g^{\sigma^{-(d-1)}}.$$

Then $(g\widetilde{\sigma})^d = h\widetilde{\sigma}^d$ and $h = aa^{-\sigma^{-d}}$. Therefore,

$$E_1((g\widetilde{\sigma})^d) = E_1(h\widetilde{\sigma}^d) = a^{-1}(h\widetilde{\sigma}^d)^e a = a^{-1}(g\widetilde{\sigma})^{de} a = x^d,$$

and if d is a proper divisor of e, then also

$$E_2((g\widetilde{\sigma})^d) = E_2(h\widetilde{\sigma}^d) = a^{-1}(h\widetilde{\sigma}^d)^{e/d} a = a^{-1}(g\widetilde{\sigma})^e a = x,$$

which completes the proof. □

Remark 3.2.6 Let $g, h \in X_{\sigma^e}$. If $g\widetilde{\sigma}$ and $h\widetilde{\sigma}$ are $\langle X_{\sigma^e}, \widetilde{\sigma}\rangle$-conjugate, then there exist $k \in X_{\sigma^e}$ and an integer i such that

$$h\widetilde{\sigma} = (k\widetilde{\sigma}^i)^{-1} g\widetilde{\sigma}(k\widetilde{\sigma}^i) = (h\widetilde{\sigma})^i (k\widetilde{\sigma}^i)^{-1} g\widetilde{\sigma}(k\widetilde{\sigma}^i)(h\widetilde{\sigma})^{-i},$$

but $(k\widetilde{\sigma}^i)(h\widetilde{\sigma})^{-i} \in X_{\sigma^e}$, so $g\widetilde{\sigma}$ and $h\widetilde{\sigma}$ are X_{σ^e}-conjugate. In particular,

$$|C_{\langle X_{\sigma^e}, \widetilde{\sigma}\rangle}(g\widetilde{\sigma})| = e|C_{X_{\sigma^e}}(g\widetilde{\sigma})|. \tag{3.3}$$

3.3 Applications

Theorem 3.1.4(ii) demonstrates that Shintani maps preserve important group theoretic data. We now exploit this by providing three applications of Shintani descent to the problem of determining maximal overgroups of elements. We continue to assume that X is a connected algebraic group, σ is a Steinberg endomorphism of X, $e > 1$, F is the Shintani map of (X, σ, e) and $\widetilde{\sigma} = \sigma|_{X_{\sigma^e}}$.

We begin with an important result [21, Theorem 2.14].

Theorem 3.3.1 *Let Y be a closed connected σ-stable subgroup of X. For all elements $g \in X_{\sigma^e}$,*

$$\operatorname{fix}(g\widetilde{\sigma}, X_{\sigma^e}/Y_{\sigma^e}) = \operatorname{fix}(F(g\widetilde{\sigma}), X_\sigma/Y_\sigma).$$

The first application naturally extends [21, Corollary 2.15].

Lemma 3.3.2 *Let Y be a closed connected σ-stable subgroup of X such that $N_{X_\sigma}(Y_\sigma) = Y_\sigma$ and $N_{X_{\sigma^e}}(Y_{\sigma^e}) = Y_{\sigma^e}$. For all $g \in X_{\sigma^e}$, the number of X_{σ^e}-conjugates of Y_{σ^e} normalised by $g\widetilde{\sigma}$ equals the number of X_σ-conjugates of Y_σ containing $F(g\widetilde{\sigma})$.*

Proof Since Y_{σ^e} is σ-stable and $N_{X_{\sigma^e}}(Y_{\sigma^e}) = Y_{\sigma^e}$, the conjugation action of $\langle X_{\sigma^e}, \widetilde{\sigma}\rangle$ on the set of X_{σ^e}-conjugates of Y_{σ^e} is equivalent to the action of $\langle X_{\sigma^e}, \widetilde{\sigma}\rangle$ on cosets of Y_{σ^e} in X_{σ^e}. Therefore, the number of X_{σ^e}-conjugates of Y_{σ^e} normalised by $g\widetilde{\sigma}$ is $\operatorname{fix}(g\widetilde{\sigma}, X_{\sigma^e}/Y_{\sigma^e})$. Similarly, the number of X_σ-conjugates of Y_σ containing $F(g\widetilde{\sigma})$ is $\operatorname{fix}(F(g\widetilde{\sigma}), X_\sigma/Y_\sigma)$. The result now follows from Theorem 3.3.1. □

The following example demonstrates a typical application of Lemma 3.3.2.

Example 3.3.3 Let $n \geqslant 2$ and let $q = q_0^e = p^f$ where $e \geqslant 2$ divides f. Let $X = \mathrm{SL}_n(\overline{\mathbb{F}}_p)$ and let $\sigma = \varphi^{f/e}$, where φ is the standard Frobenius endomorphism $(a_{ij}) \mapsto (a_{ij}^p)$ of X, with respect to some fixed basis $\mathcal{B} = (u_1, \dots, u_n)$ for $\overline{\mathbb{F}}_p^n$. Let F be the Shintani map of (X, σ, e). Note that $X_\sigma = \mathrm{SL}_n(q_0)$ and $X_{\sigma^e} = \mathrm{SL}_n(q)$.

Let $1 \leqslant k < n$. We may fix a σ-stable maximal parabolic subgroup $Y \leqslant X$ of type P_k; for example, let Y be the stabiliser in X of the subspace $\langle u_1, \dots, u_k\rangle$. In particular, Y is a closed connected subgroup of X. Moreover, $N_{X_\sigma}(Y_\sigma) = Y_\sigma$ and $N_{X_{\sigma^e}}(Y_{\sigma^e}) = Y_{\sigma^e}$, so we are in a position to apply Lemma 3.3.2.

Let $g \in X_{\sigma^e}$. By Lemma 3.3.2, the number of $\mathrm{SL}_n(q)$-conjugates of Y_{σ^e} normalised by $g\widetilde{\sigma}$ equals the number of $\mathrm{SL}_n(q_0)$-conjugates of Y_σ containing $F(g\widetilde{\sigma})$.

There is a unique $\mathrm{SL}_n(q)$-class of maximal subgroups of $G = \langle \mathrm{SL}_n(q), \widetilde{\sigma}\rangle$ of type P_k and this class is represented by $H = \langle Y_{\sigma^e}, \widetilde{\sigma}\rangle$ (see, for example, [50, Proposition 4.1.17]). In addition, for each $x \in \mathrm{SL}_n(q)$, the element $g\widetilde{\sigma}$ is contained in H^x if and only if $g\widetilde{\sigma}$ normalises $Y_{\sigma^e}^x$. Therefore, the number of G-conjugates of H containing $g\widetilde{\sigma}$ equals the number of $\mathrm{SL}_n(q_0)$-conjugates of Y_σ containing $F(g\widetilde{\sigma})$.

Example 3.3.3 highlights the key idea of Shintani descent: we can deduce information about $g\widetilde{\sigma}$ from information about $F(g\widetilde{\sigma})$.

Our second application is a minor generalisation of [21, Proposition 2.16(i)]. Here we write $\widetilde{G} = X_{\sigma^e}{:}\langle\widetilde{\sigma}\rangle$.

Lemma 3.3.4 *Let $g \in \widetilde{G}$ and let $H \leqslant \widetilde{G}$. Then $g\widetilde{\sigma}$ is contained in at most $|C_{X_\sigma}(F(g\widetilde{\sigma}))|$ distinct $\widetilde{G}$-conjugates of H.*

Proof By Lemma 2.1.2, the number of $\widetilde{G}$-conjugates of H that contain $g\widetilde{\sigma}$ is

$$N = \frac{|(g\widetilde{\sigma})^{\widetilde{G}} \cap H|}{|(g\widetilde{\sigma})^{\widetilde{G}}|} \cdot \frac{|\widetilde{G}|}{|N_{\widetilde{G}}(H)|} = \frac{|(g\widetilde{\sigma})^{\widetilde{G}} \cap H||C_{\widetilde{G}}(g\widetilde{\sigma})|}{|N_{\widetilde{G}}(H)|}.$$

First note that $(g\widetilde{\sigma})^{\widetilde{G}} \subseteq X_{\sigma^e} g\widetilde{\sigma}$, and for $0 \leqslant i < e$, the cosets $(X_{\sigma^e} \cap H)(g\widetilde{\sigma})^i$ in H are distinct. Therefore, $|(g\widetilde{\sigma})^{\widetilde{G}} \cap H| \leqslant |H|/e$. Next, by (3.3) and Theorem 3.1.4(ii),

$$|C_{\widetilde{G}}(g\widetilde{\sigma})| = |C_{X_{\sigma^e}}(g\widetilde{\sigma})|e = |C_{X_\sigma}(F(g\widetilde{\sigma}))|e.$$

Together these observations give

$$N \leqslant \frac{|H||C_{X_\sigma}(F(g\widetilde{\sigma}))|e}{e|N_{\widetilde{G}}(H)|} \leqslant |C_{X_\sigma}(F(g\widetilde{\sigma}))|.$$

This completes the proof. □

The third application is based on [21, Proposition 2.16(ii)] and is more specialised than the previous two. To state this result, we need to fix some notation.

Let X be a simple classical algebraic group of adjoint type. Let $\sigma = \rho\varphi^i$, where φ is a standard Frobenius endomorphism of X and one of the following holds

(i) ρ is trivial,
(ii) $X = \mathsf{A}_m$ and ρ is the standard graph automorphism γ
(iii) $X = \mathsf{D}_m$ and ρ is the standard graph automorphism r

(see Definition 2.5.1). Let $x \in X_\sigma$ have odd order and act on the natural module V for X_σ as $A_1 \oplus \cdots \oplus A_k$ centralising a decomposition $V = U_1 \oplus \cdots \oplus U_k$, where for each $1 \leqslant i \leqslant k$, one of the following holds

(i) A_i is irreducible on the d_i-space U_i
(ii) $(X, \rho) \neq (\mathsf{A}_m, 1)$ and $A_i = B_i \oplus B_i^*$ stabilises $U_i = W_i \oplus W_i^*$, where $\{W_i, W_i^*\}$ is a dual pair of totally singular d_i-spaces and B_i and B_i^* are irreducible on W_i and W_i^*, respectively, and are not similar.

Lemma 3.3.5 *Let $g \in X_{\sigma^e}$ such that $F(g\widetilde{\sigma})$ is the element x defined above. If e is prime and $(d_i, d_j) = 1$ when $i \neq j$, then the number of X_{σ^e}-conjugates of X_σ normalised by $g\widetilde{\sigma}$ is at most e^k.*

Proof Write $\widetilde{G} = X_{\sigma^e}{:}\langle\widetilde{\sigma}\rangle$, and write $H = X_\sigma$ and $\widetilde{H} = N_{\widetilde{G}}(H)$, noting that $\widetilde{H} = H \times \langle\widetilde{\sigma}\rangle$ since H is adjoint. For a subset $S \subseteq \overline{\mathbb{F}}_q^{\times}$, let S^* be $S^{-\sigma}$ if H is a unitary group and S^{-1} otherwise. The restrictions on $F(g\widetilde{\sigma})$ in the statement imply that the eigenvalue multiset (over $\overline{\mathbb{F}}_p$) of $F(g\widetilde{\sigma})$ is $S_1 \cup \cdots \cup S_k$ where S_i is either Λ_i or $\Lambda_i \cup \Lambda_i^*$ where $\Lambda_i = \{\lambda_i, \ldots, \lambda_i^{q_0^{d_i-1}}\}$, and $\Lambda_i \neq \Lambda_i^*$ in the latter case.

Let $h\widetilde{\sigma} \in \widetilde{H}$ be $\widetilde{G}$-conjugate to $g\widetilde{\sigma}$. Then $F(h\widetilde{\sigma})$ is X_σ-conjugate to $F(g\widetilde{\sigma})$. Let the eigenvalue multiset of $h \in H$ be $\{\alpha_1, \ldots, \alpha_n\}$. Therefore, the eigenvalue multiset of $F(h\widetilde{\sigma})$ is the eigenvalue multiset of $(h\widetilde{\sigma})^e = h^e$, which is $\{\alpha_1^e, \dots, \alpha_n^e\}$. Therefore, without loss of generality, $\alpha_i^e = \lambda_i$ for each $1 \leqslant i \leqslant k$. Now note that $\alpha_1, \dots, \alpha_k$ determine all of the eigenvalues of h. Thus, there are e^k choices for the eigenvalues of h and consequently e^k choices for h and, hence, $h\widetilde{\sigma}$ up to H-conjugacy. Therefore, $(g\widetilde{\sigma})^{\widetilde{G}} \cap \widetilde{H}$ splits into e^k H-classes. Since $(d_i, d_j) = 1$ for $i \neq j$, we know that h stabilises the same type of decomposition of $\mathbb{F}_{q_0}^n$ as g, acting irreducibly on the corresponding summands. Therefore, $|C_{X_g}(h\widetilde{\sigma})| = |C_{X_\sigma}(h)| = |C_{X_\sigma}(F(g\widetilde{\sigma}))|$. Consequently, the H-classes into which $g\widetilde{\sigma}^{\widetilde{G}} \cap \widetilde{H}$ splits have size $|F(g\widetilde{\sigma})^H|$.

By Lemma 2.1.2, the number of $\widetilde{G}$-conjugates of $\widetilde{H}$ which contain $g\widetilde{\sigma}$ is

$$\frac{|(g\widetilde{\sigma})^{\widetilde{G}} \cap \widetilde{H}|}{|(g\widetilde{\sigma})^{\widetilde{G}}|}\frac{|\widetilde{G}|}{|\widetilde{H}|} \leqslant \frac{e^k|F(g\widetilde{\sigma})^{X_\sigma}||C_{\widetilde{G}}(g\widetilde{\sigma})|}{|\widetilde{H}|} = \frac{e^k|X_\sigma||C_{\widetilde{G}}(g\widetilde{\sigma})|}{|\widetilde{H}||C_{X_\sigma}(F(g\widetilde{\sigma}))|} \leqslant e^k.$$

This completes the proof. □

3.4 Generalisation

We saw in Example 3.1.5 that there are situations that we will encounter in the proof of our main theorems where Theorem 3.1.4 alone does not allow us to identify and work with elements in the relevant coset of the almost simple group. These are the situations that we will encounter in Sects. 5.4.2, 5.5.2, and 6.4.2. We now present a new result that allows us to handle these cases.

Let X be a connected algebraic group, σ a Steinberg endomorphism of X and $e > 1$. Further, let ρ be an automorphism (of algebraic groups) of X.

Suppose that we wish to identify an element in the coset $X_{\rho\sigma^e}\widetilde{\sigma}$ (compare with Example 3.1.5). The rough idea of Lemma 3.4.1 is that while we cannot find elements $x \in X_{\rho\sigma}$ and $g \in X_{\rho\sigma^e}$ such that $(g\widetilde{\sigma})^e$ is X-conjugate to x, if we insist that x is contained in $(C_X(\rho)^\circ)_{\rho\sigma} \leqslant X_{\rho\sigma}$, then there does exist $g \in X_{\rho\sigma^e}$ such that $(g\widetilde{\sigma})^e$ is X-conjugate to $x\widetilde{\rho}^{-1}$.

Lemma 3.4.1 *Let Z be a closed connected σ-stable subgroup of X contained in $C_X(\rho)$. Let $G = X_{\rho\sigma^e}{:}\langle\widetilde{\rho}, \widetilde{\sigma}\rangle$ where $\widetilde{\sigma} = \sigma|_{X_{\rho\sigma^e}}$ and $\widetilde{\rho} = \rho|_{X_{\rho\sigma^e}}$. Let $x \in Z_\sigma \leqslant X_{\rho\sigma^e}$.*

(i) *There exists $g \in Z_{\sigma^e} \leqslant X_{\rho\sigma^e}$ such that $(g\widetilde{\sigma})^e$ and $x\widetilde{\rho}^{-1}$ are X-conjugate elements of G.*
(ii) *Moreover, if there exists a positive integer d such that $(\rho\sigma^e)^d = \sigma^{ed}$, then for any g as in (i) the following hold.*

  (a) *Let $H \leqslant \langle X_{\rho\sigma^e}, \widetilde{\sigma}\rangle$. Then the number of $X_{\rho\sigma^e}$-conjugates of H that contain $g\widetilde{\sigma}$ is at most $|C_{X_\sigma}(x^d)|$.*

(b) *Let Y be a closed connected σ-stable subgroup X such that $N_{X_\sigma}(Y_\sigma) = Y_\sigma$ and $N_{X_{\sigma^{de}}}(Y_{\sigma^{de}}) = Y_{\sigma^{de}}$. Then the number of $X_{\sigma^{de}}$-conjugates of $Y_{\sigma^{de}}$ normalised by $g\widetilde{\sigma}$ equals the number of X_σ-conjugates of Y_σ containing x^d.*

Proof Let F be the Shintani map of (Z, σ, e) and fix $x \in Z_\sigma$. Let $\widehat{\sigma} = \sigma|_{Z_{\sigma^e}}$, noting that $\widehat{\sigma}^e = 1$. By Theorem 3.1.4 applied to F, there exists $g \in Z_{\sigma^e}$ such that

$$a^{-1}(g\widehat{\sigma})^e a = a^{-1}(gg^{\sigma^{e-1}}g^{\sigma^{e-2}}\cdots g^{\sigma})a = x$$

as elements of $Z_{\sigma^e}{:}\langle\widehat{\sigma}\rangle$, where $a \in Z \leqslant X$ satisfies $aa^{-\sigma^{-1}} = g$. Now $g \in Z_{\sigma^e} \leqslant X_{\rho\sigma^e}$ and $\widetilde{\sigma}^e = \widetilde{\rho}^{-1}$ as an element of $G = X_{\rho\sigma^e}{:}\langle\widetilde{\sigma}, \widetilde{\rho}\rangle$. Therefore, as elements of G,

$$a^{-1}(g\widetilde{\sigma})^e a = a^{-1}(gg^{\sigma^{e-1}}g^{\sigma^{e-2}}\cdots g^{\sigma})\widetilde{\sigma}^e a = a^{-1}(gg^{\sigma^{e-1}}g^{\sigma^{e-2}}\cdots g^{\sigma})\widetilde{\rho}^{-1}a = x\widetilde{\rho}^{-1}.$$

This proves part (i).

Now turn to part (ii) and assume that $(\rho\sigma^e)^d = \sigma^{ed}$. Let E be the Shintani map of (X, σ, de), recording that $Z_\sigma \leqslant X_\sigma$ and $X_{\rho\sigma^e} \leqslant X_{(\rho\sigma^e)^d} = X_{\sigma^{de}}$. Write $\overline{\sigma} = \sigma|_{X_{\sigma^{de}}}$. Since $\overline{\sigma}|_{X_{\rho\sigma^e}} = \widetilde{\sigma}$ and $|\overline{\sigma}| = de = |\widetilde{\sigma}|$, we can consider $\langle X_{\rho\sigma^e}, \widetilde{\sigma}\rangle$ as a subgroup of $\langle X_{\sigma^{de}}, \overline{\sigma}\rangle$, where we identify $\widetilde{\sigma}$ with $\overline{\sigma}$. Consequently, $E(g\widetilde{\sigma}) = a^{-1}(g\widetilde{\sigma})^{de}a = x^d$. By Lemma 3.3.4, if $H \leqslant \langle X_{\rho\sigma^e}, \widetilde{\sigma}\rangle$, then the number of $X_{\sigma^{de}}$-conjugates of H containing $g\widetilde{\sigma}$ is at most $|C_{X_\sigma}(x^d)|$, which implies (a). If Y is a connected σ-stable subgroup X such that $N_{X_\sigma}(Y_\sigma) = Y_\sigma$ and $N_{X_{\sigma^{de}}}(Y_{\sigma^{de}}) = Y_{\sigma^{de}}$, then Lemma 3.2.1 implies that the number of $X_{\sigma^{de}}$-conjugates of $Y_{\sigma^{de}}$ normalised by $g\widetilde{\sigma}$ equals the number of X_σ-conjugates of Y_σ containing x^d, as claimed in (b). □

Example 3.4.2 This continues Example 3.1.5. Let $e \geqslant 3$ be odd, $m \geqslant 4$ and $q = 2^e$. Write $X = \Omega_{2m}(\overline{\mathbb{F}}_2)$ and recall the standard Frobenius endomorphism φ and the involutory automorphism r. The existing Shintani descent methods did not provide information about the coset $\Omega^+_{2m}(q)r\varphi$ of $\langle\Omega^+_{2m}(q), r\varphi\rangle$. We now use Lemma 3.4.1 to overcome this obstacle.

Let $Z \cong \Omega_{2m-2}(\overline{\mathbb{F}}_2)$ be the subgroup of X that centralises $\langle e_1, \ldots, f_{m-1}\rangle \perp \langle e_m, f_m\rangle$ and acts trivially on the second summand. Evidently $Z \leqslant C_X(r)$. Therefore, Lemma 3.4.1(i) implies that for all $x \in Z_{r\varphi} \cong \Omega^+_{2m-2}(2)$, there exists $g \in X_{r(r\varphi)^e} = \Omega^+_{2m}(q)$ such that $(gr\varphi)^e$ is X-conjugate to xr. Crucially, parts (a) and (b) of Lemma 3.4.1(ii) translate information about x into information about $gr\varphi$, so, in this way, we can select and work with elements in the coset $\Omega^+_{2m}(q)r\varphi$.

Chapter 4
Fixed Point Ratios

This chapter presents upper bounds on fixed point ratios that we will use as part of the probabilistic method we described in Sect. 2.1. Much is known about fixed point ratios for primitive actions of almost simple groups. One reason for this is the important applications these bounds have to a diverse range of problems, such as monodromy groups and base sizes of permutation groups, via probabilistic methods (see [18]).

For groups of Lie type, the most general such bound is [54, Theorem 1] of Liebeck and Saxl, which establishes that

$$\mathrm{fpr}(x, G/H) \leqslant \frac{4}{3q} \tag{4.1}$$

for any almost simple group of Lie type G over $\mathbb{F}_q$, core-free maximal subgroup H of G and nontrivial element $x \in G$, unless $\mathrm{soc}(G) \in \{\mathrm{PSL}_2(q)\} \cup \{\mathrm{PSL}_4(2), \mathrm{PSp}_4(3)\}$. This bound is essentially best possible, since $\mathrm{fpr}(x, G/H)$ is roughly q^{-1} when q is odd, $G = \mathrm{PGL}_n(q)$, H is a maximal P_1 parabolic subgroup (the stabiliser of a 1-space of $\mathbb{F}_q^n$) and x is a reflection (see [18, Example 1.5]).

Let G be an almost simple classical group. A maximal subgroup $H \leqslant G$ is a *subspace subgroup* if $H \cap \mathrm{soc}(G)$ acts reducibly on the natural module for $\mathrm{soc}(G)$ or if $\mathrm{soc}(G)$ is $\mathrm{Sp}_n(2^f)$ and $H \cap \mathrm{soc}(G) = \mathrm{O}_n^{\pm}(2^f)$, and H is a *nonsubspace subgroup* otherwise. In Sect. 4.1 we record and prove bounds on fixed point ratios for subspace subgroups.

Notice that the bound in (4.1) does not depend on the element x. In the sequence of papers [14–17], Burness establishes an upper bound on $\mathrm{fpr}(x, G/H)$ depending on x when $H \leqslant G$ is nonsubspace and $x \in G$ has prime order. We present and apply this result in Sect. 4.2, before giving some tighter bounds on the fixed point ratios for nonsubspace actions of almost simple unitary groups in low dimensions.

S. Harper, *The Spread of Almost Simple Classical Groups*, Lecture Notes in Mathematics 2286, https://doi.org/10.1007/978-3-030-74100-6_4

4.1 Subspace Actions

For this section, write $q = p^f$. The first theorem combines several results of Guralnick and Kantor [38, Section 3]. Here $d = 2$ if $\mathrm{soc}(G) = \mathrm{PSU}_n(q)$ and $d = 1$ otherwise, so $\mathbb{F}_{q^d}^n$ is the natural module for $\mathrm{GL}_n(q)$, $\mathrm{Sp}_n(q)$, $\mathrm{O}_n^\varepsilon(q)$ and $\mathrm{GU}_n(q)$.

Theorem 4.1.1 *Let $G \leqslant \mathrm{P\Gamma L}(V)$ be an almost simple classical group with natural module $V = F^n$ where $F = \mathbb{F}_{q^d}$. Assume that $n \geqslant 6$. Let $H \leqslant G$ be a reducible maximal subgroup, stabilising a subspace $0 < U < V$ of dimension k and Witt index l. Let $1 \neq x \in G$. Let m, a, b, c be the parameters defined in Fig. 4.1.*

(i) *If* $\mathrm{soc}(G) = \mathrm{PSL}_n(q)$, *then*

$$\mathrm{fpr}(x, G/H) \leqslant 2|F|^{-\min\{k,\, n-k\}}.$$

(ii) *If* $\mathrm{soc}(G) \neq \mathrm{PSL}_n(q)$ *and U is nondegenerate, then*

$$\mathrm{fpr}(x, G/H) \leqslant 2|F|^{-m+a} + |F|^{-m+b} + |F|^{-l} + |F|^{-n+k}.$$

(iii) *If* $\mathrm{soc}(G) \neq \mathrm{PSL}_n(q)$ *and U is totally singular, then*

$$\mathrm{fpr}(x, G/H) \leqslant 2|F|^{-m+c} + |F|^{-\frac{m}{d}+\frac{b}{d}} + |F|^{-k}.$$

Frohardt and Magaard established upper and lower bounds on the fixed point ratio of an element x of an almost simple classical group G on an appropriate set of k-spaces of V that depend on G and k but also take into account the element x [34, Theorems 1–6]. In particular, these bounds depend on the following invariant.

Notation 4.1.2 For $x \in \mathrm{PGL}_n(q)$, let $\hat{x}$ be a preimage of x in $\mathrm{GL}_n(q)$ and define $\nu(x)$ as the codimension of the largest eigenspace of $\hat{x}$ on $V \otimes_{\mathbb{F}_q} \overline{\mathbb{F}}_p$.

For example, if $G = \mathrm{P\Omega}_n^\varepsilon(q)$ and $H \leqslant G$ is the stabiliser of a nondegenerate k-space, where $k < \frac{n}{2}$, then for $x \in G$ with $s = \nu(x) < \frac{n}{2k}$, [34, Theorem 2] gives

$$q^{-sk} - 3q^{-(n-1)/2} < \mathrm{fpr}(x, G/H) < q^{-sk} + 200q^{-(n-1)/2}.$$

Fig. 4.1 The values $m, a, b,$ c in Theorem 4.1.1

$\mathrm{soc}(G)$		a	b	c
$\mathrm{PSp}_{2m}(q)$	q even	1	0	1
	q odd	2	0	1
$\Omega_{2m+1}(q)$		1	0	1
$\mathrm{P\Omega}_{2m}^\varepsilon(q)$	$\varepsilon = +$	2	1	2
	$\varepsilon = -$	2	0	1
$\mathrm{PSU}_n(q)$	$n = 2m$	2	$\frac{1}{2}$	1
	$n = 2m+1$	1	$-\frac{1}{2}$	0

As in [43], the constants in these upper bounds are too large for our application, so we derive our own bounds for the particular cases we will require. Our bounds have no restriction on s in terms of k.

In part (i) of the statement of Proposition 4.1.3, if q is even, then the nonsingular 1-space in question is degenerate and has a stabiliser of type $\mathrm{Sp}_{2m-2}(q)$ (see [50, Proposition 4.1.7]).

Proposition 4.1.3 *Let $G = \mathrm{PO}^{\varepsilon}_{2m}(q)$ where $m \geqslant 4$. Let $x \in G$ have prime order and $\nu(x) = s$.*

(i) *If $H \leqslant G$ is the stabiliser of a nonsingular 1-space, then*

$$\mathrm{fpr}(x, G/H) \leqslant \frac{1}{q^s} + \frac{1}{q^{2m-s}} + \frac{2}{q^m - \varepsilon}.$$

(ii) *If $H \leqslant G$ is the stabiliser of a nondegenerate 2-space, then*

$$\mathrm{fpr}(x, G/H) \leqslant \frac{1}{q^{2s}} + \frac{1}{q^{2m-2s}} + \frac{1}{q^{m-1} - 1} + \frac{4}{q^{2m-3}}.$$

Proof Let r be the order of x. If x is not contained in a G-conjugate of H, then $\mathrm{fpr}(x, G/H) = 0$. Therefore, assume that $x \in H$. Let $V = \mathbb{F}_q^{2m}$.

Proof of Part (i) Write $H = G_{\langle u \rangle}$ and $U = \langle u \rangle^{\perp}$. We consider four cases depending on r and p.

Case 1: $r \notin \{2, p\}$ In this case, x is a semisimple element of odd prime order. By [20, Proposition 3.5.4], x is G-conjugate to an element that lifts to a block diagonal matrix $[A_1^{a_1}, \dots, A_t^{a_t}, I_e]$ centralising a decomposition $V = V_1^{a_1} \perp \cdots \perp V_t^{a_t} \perp W$ where, for some even k, each V_j is a nondegenerate k-space and W is the (nondegenerate) 1-eigenspace of x. Moreover, either each matrix A_j acts irreducibly on V_j or each matrix A_j centralises the decomposition $V_j = U_j \oplus U_j^*$, where U_j and U_j^* are totally singular subspaces on which A_j acts irreducibly. We may assume that the submodules V_j are pairwise nonisomorphic.

Since $x \in H$, we deduce that x fixes u. Therefore, $e > 0$ and on U the element x acts as $[M_1, \dots, M_d, I_{e-1}]$. Therefore, [20, Lemma 3.5.3] implies that $x^G \cap H = x^H$. Moreover, from the centraliser orders in [20, Appendix B], writing $e = 2l$, we obtain

$$\frac{|x^G \cap H|}{|x^G|} = \frac{|H|}{|G|}\frac{|C_G(x)|}{|C_H(x)|} \leqslant \frac{(2, q-1)}{q^{m-1}(q^m - \varepsilon)}\frac{q^{l-1}(q^l + 1)}{(2, q-1)} \leqslant \frac{1}{q^{2m-2l}} + \frac{1}{q^m - \varepsilon}.$$

Since $2l$ is the dimension of the 1-eigenspace of x, we know that $2m - 2l \geqslant s$. The result now follows in this case.

Case 2: $r = p = 2$ Here x is a unipotent involution and we adopt the notation of Aschbacher and Seitz [4]. Since $p = 2$, the subgroup H is the centraliser in G

of a b_1 involution, and we may write $H \cong \langle b_1 \rangle \times \mathrm{Sp}_{2m-2}(q)$. Now an element $x = (y, z) \in H$, where $y \in \langle b_1 \rangle$ and $z \in \mathrm{Sp}_{2m-2}(q)$, embeds as the block diagonal element $[y, z]$ in G. Therefore, in light of [20, Lemma 3.4.14], it is straightforward to determine how $x^G \cap H$ splits into H-classes. For example, if $x = b_s$ for odd $s \geqslant 3$, then $x^G \cap H$ is the union of $x_1^H \cup x_2^H \cup x_3^H$ where x_1, x_2 and x_3 are the elements (I_2, b_s), (b_1, a_{s-1}) and (b_1, c_{s-1}) of H. Therefore, using the centraliser orders that are given in [20, Appendix B], we deduce that

$$\begin{aligned}\mathrm{fpr}(x, G/H) &= \frac{|H|}{|G|}\sum_{i=1}^{3}\frac{|C_G(x_i)|}{|C_H(x_i)|} = \frac{q^{s-1}(q^{2m-2s}-1)+1+(q^{s-1}-1)}{q^{m-1}(q^m-\varepsilon)} \\ &= \frac{q^{m-s}}{q^m-\varepsilon} \leqslant \frac{1}{q^s} + \frac{1}{q^m-\varepsilon}.\end{aligned}$$

In the remaining cases, $x^G \cap H$ splits into H-classes in the following ways

$b_1^G \cap H = (b_1, I_{n-2})^H \cup (I_2, b_1)^H$
$b_m^G \cap H = (b_1, a_{m-1})^H \cup (b_1, c_{m-1})^H$, where m is necessarily odd
$a_s^G \cap H = (I_2, a_s)^H$, if $s < m$
$c_s^G \cap H = (I_2, c_s)^H \cup (b_1, b_{s-1})^H$, if $s < m$
$c_m^G \cap H = (b_1, b_{m-1})^H$, where m is necessarily even,

noting that x does not have type a_m (when m is even), since $a_m^G \cap H$ is empty. In each case, we can verify the claimed bound as above.

Case 3: $r = 2$ and $p > 2$ The G-classes of semisimple involutions are described in detail in [20, Section 3.5.2]. Since $x \in H$ we may deduce that x has type t_i, t_i' or γ_i for some i, in the notation of [20, 37]. (In particular, [20, Table B.9] makes clear that involutions arising from matrices of order four do not stabilise nondegenerate 1-spaces.) Said otherwise, x lifts to an involution $-I_a \perp I_b$ centralising a decomposition $U_1 \perp U_2$ where U_1 and U_2 are nondegenerate a- and b-spaces. Therefore, either x fixes u and acts as $-I_a \perp I_{b-1}$ on U, or x negates u and acts as $-I_{a-1} \perp I_b$ on U. Therefore, $x^G \cap H = x_1^H \cup x_2^H$ where x_1 and x_2 correspond to the two possible actions of x on u. Consequently,

$$\frac{|x^G \cap H|}{|x^G|} = \frac{|H|}{|G|}\left(\frac{|C_G(x_1)|}{|C_H(x_1)|} + \frac{|C_G(x_2)|}{|C_H(x_2)|}\right).$$

Assume that $a = 2k$ and $b = 2l$; the case where a and b are odd is very similar. From the centraliser orders in [20, Appendix B] we can compute that

$$\frac{|C_G(x_1)|}{|C_H(x_1)|} \leqslant \frac{1}{2}q^{l-1}(q^l+1) \quad \text{and} \quad \frac{|C_G(x_2)|}{|C_H(x_2)|} \leqslant \frac{1}{2}q^{k-1}(q^k+1).$$

Therefore,

$$\frac{|H|}{|G|}\left(\frac{|C_G(x_1)|}{|C_H(x_1)|}+\frac{|C_G(x_2)|}{|C_H(x_2)|}\right)\leqslant \frac{q^{l-1}(q^l+1)+q^{k-1}(q^k+1)}{q^{m-1}(q^m-\varepsilon)}\leqslant \frac{1}{q^{2k}}+\frac{1}{q^{2l}}+\frac{1}{q^m-\varepsilon}.$$

Since $\{2k,2l\}=\{s,2m-s\}$, we have verified the result in this case.

Case 4: $r=p>2$ In this case, x is a unipotent element of odd prime order, so, by [20, Proposition 3.5.12], x is G-conjugate to an element that lifts to a matrix with Jordan form $[J_p^{a_p},\dots,J_2^{a_2},J_1^{a_1}]$ where $\sum_{i=1}^p ia_i=2m$. Indeed, the conjugacy class x^G is characterised by this Jordan form together with a sequence $(\delta_1,\delta_3,\dots,\delta_p)$ in $\{\square,\boxtimes\}$ that satisfies $\delta_1\delta_3\cdots\delta_p=D(Q)$, where Q is the form defining G.

Note that $V=\langle u\rangle\perp U$ since p is odd. Since $x\in H$ and the only eigenvalue of x is 1, the vector u is fixed by x. Since the 1-eigenspace of J_i is totally singular when $i>1$, we deduce that $a_1>0$ and x acts on U as an element whose Jordan form is $[J_p^{a_p},\dots,J_2^{a_2},J_1^{a_1-1}]$. Moreover, the corresponding sequence of discriminants for the element $x|_U$ is $(\delta_1\delta,\delta_3,\dots,\delta_p)$, where $\delta=D(Q|_{\langle u\rangle})$. By [20, Proposition 3.5.12], this completely determines the H-class of x. Therefore, $x^G\cap H=x^H$. Using the centraliser orders in [20, Appendix B], noting that $s=2m-\sum_{j\geqslant 1}a_j$, we have

$$\mathrm{fpr}(x,G/H)\leqslant \frac{|H|}{|G|}\frac{|C_G(x)|}{|C_H(x)|}=\frac{q^{2m-s-a_1}}{q^{m-1}(q^m-\varepsilon)}\frac{|\mathrm{O}_{a_1}^{\varepsilon_1}(q)|}{|\mathrm{O}_{a_1-1}^{\varepsilon_1'}(q)|}\leqslant \frac{1}{q^s}+\frac{1}{q^m-\varepsilon}.$$

This completes the proof of part (i).

Proof of Part (ii) We proceed in the same way as for part (i).

Case 1: $r\notin\{2,p\}$ In this case, x is a semisimple element of odd prime order. By [20, Proposition 3.5.4], x is G-conjugate to an element that lifts to a block diagonal matrix $[A_1^{a_1},\dots,A_t^{a_t},I_e]$ centralising a decomposition $V=V_1^{a_1}\perp\cdots\perp V_t^{a_t}\perp W$ where, for some even k, each V_j is a nondegenerate k-space and W is the (nondegenerate) 1-eigenspace of x. Moreover, either each matrix A_j acts irreducibly on V_j or each matrix A_j centralises the decomposition $V_j=U_j\oplus U_j^*$, where U_j and U_j^* are totally singular subspaces on which A_j acts irreducibly. We may assume that the submodules V_j are pairwise nonisomorphic.

Let us now determine how $x^G\cap H$ splits into H-classes. Let $h\in H$ be G-conjugate to x. Then h lifts to $(A,B)\in\mathrm{O}_2^{\varepsilon_1}(q)\times\mathrm{O}_{2m-2}^{\varepsilon_2}(q)$. If $A=I_2$, then $e\geqslant 2$ and h is H-conjugate to x_0, an element lifting to $(I_2,[A_1^{a_1},\dots,A_t^{a_t},I_{e-2}])$. If $A\neq I_2$, then let $\lambda\in\overline{\mathbb{F}}_q$ be a nontrivial eigenvalue of A. Then λ is an eigenvalue of A_j for some j. Since the set of eigenvalues of A is closed under the map $\mu\mapsto\mu^q$, we deduce that $k=2$ and $A=A_j$. Therefore, h is H-conjugate to x_j, an element lifting to $(A_j,[A_1^{a_1},\dots,A_j^{a_j-1},\dots,A_t^{a_t},I_e])$.

This information is enough to determine how $x^G \cap H$ splits into H-classes. If $k > 2$, then $e > 0$ and $x^G \cap H = x_0^H$. If $k = 2$, then, writing $e = 2a_0$, we have

$$x^G \cap H = \bigcup_{\substack{0 \leqslant j \leqslant t \\ a_j > 0}} x_j^H$$

We now use this information about $x^G \cap H$ to find an upper bound on $\mathrm{fpr}(x, G/H)$. First note that

$$\frac{|H|}{|G|} = \frac{|\,\mathrm{O}_2^{\varepsilon_1}(q)||\,\mathrm{O}_{2m-2}^{\varepsilon_2}(q)|}{|\,\mathrm{O}_{2m}^{\varepsilon}(q)|} = \frac{2(q-\varepsilon_1)}{q^{2m-2}(q^{m-1}+\varepsilon_2)(q^m-\varepsilon)}.$$

Similarly, if $e = 2a_0 > 0$, then

$$\frac{|C_G(x)|}{|C_H(x_0)|} \leqslant \frac{q^{e-2}(q^{a_0-1}+1)(q^{a_0}+1)}{2(q-\varepsilon_1)}.$$

Now assume that $k = 2$. Let $\eta = +$ if r divides $q-1$ and let $\eta = -$ otherwise (when r necessarily divides $q+1$). Then for all $1 \leqslant j \leqslant t$ such that $a_j > 0$ we have

$$\frac{|C_G(x)|}{|C_H(x_j)|} = \frac{|\,\mathrm{GL}_{a_j}^{\eta}(q)|}{|\,\mathrm{GL}_1^{\varepsilon_1}(q)||\,\mathrm{GL}_{a_j-1}^{\eta}(q)|} \leqslant \frac{q^{a_j-1}(q^{a_j}+1)}{q-\varepsilon_1}$$

Now,

$$\mathrm{fpr}(x, G/H) = \frac{|H|}{|G|} \sum_{\substack{0 \leqslant j \leqslant t \\ a_j > 0}} \frac{|C_G(x)|}{|C_H(x_j)|}.$$

Therefore, with the above bounds, we maximise our upper bound on $\mathrm{fpr}(x, G/H)$ when $a_j = 0$ for all $j \geqslant 2$. In this case, $a_0 + a_1 = m$ and $s = 2a_1 = 2m - e$. Therefore,

$$\begin{aligned}
\mathrm{fpr}(x, G/H) &\leqslant \frac{2(q-\varepsilon_1)}{q^{2m-2}(q^{m-1}+\varepsilon_2)(q^m-\varepsilon)} \frac{q^{e-2}(q^{a_0-1}+1)(q^{a_0}+1)}{2(q-\varepsilon_1)} \\
&\quad + \frac{2(q-\varepsilon_1)}{q^{2m-2}(q^{m-1}+\varepsilon_2)(q^m-\varepsilon)} \frac{q^{a_1-1}(q^{a_1}+1)}{q-\varepsilon_1} \\
&\leqslant \frac{1}{q^{2s}} + \frac{4}{q^{2m-3}} + \frac{1}{q^{m-1}-1}.
\end{aligned}$$

Case 2: $r = p = 2$ and Case 3 $r = 2$ and $p > 2$ In these case, x is an involution and we proceed exactly as described in part (i); we omit the details.

Case 4: $r = p > 2$ In this case, the G-conjugacy class of x is determined by the Jordan form $[J_p^{a_p}, \ldots, J_2^{a_2}, J_1^{a_1}]$ where $\sum_{i=1}^{p} ia_i = 2m$ and a sequence $(\delta_1, \delta_3, \ldots, \delta_p)$ in $\{\square, \boxtimes\}$ where $\delta_1\delta_3 \cdots \delta_p = D(Q)$. Let $x = (y, z) \in \mathrm{O}_2^{\varepsilon_1}(q) \times \mathrm{O}_{2m-2}^{\varepsilon_2}(q)$. Since Jordan blocks of even size occur with even multiplicity in orthogonal groups in odd characteristic, we deduce that y is trivial and z has Jordan form $[J_p^{a_p}, \ldots, J_2^{a_2}, J_1^{a_1-2}]$ and sequence of discriminants $(\delta_1\delta, \delta_3, \ldots, \delta_p)$ where $\delta = D(Q|_U)$. Therefore, as in part (i), $x^G \cap H = x^H$ and the result again follows from the centraliser orders in [20, Appendix B]. We have completed the proof. □

4.2 Nonsubspace Actions

We now turn to fixed point ratios for nonsubspace actions of classical groups, which, in general, are smaller than fixed point ratios for subspace actions. Building on work of Liebeck and Shalev [55], the following general theorem was established by Burness in [14, Theorem 1] (see [14, Definition 2] for a precise definition of the *dimension of the natural module*).

Theorem 4.2.1 *Let G be an almost simple classical group such that the natural module of* $\mathrm{soc}(G)$ *is n-dimensional. If $H \leqslant G$ is a maximal nonsubspace subgroup and $x \in G$ has prime order, then* $\mathrm{fpr}(x, G/H) < |x^G|^{-\frac{1}{2}+\frac{1}{n}+\iota}$ *for ι as in [14, Table 1].*

In the statement of Theorem 4.2.1, for most subgroups $H \leqslant G$ the parameter ι is simply 0, and whenever $n \geqslant 10$ we have $\iota \leqslant \frac{1}{n-2}$. Theorem 4.2.1 is essentially best possible. For example, if $G = \mathrm{PSL}_n(q_0^2)$ and $H = \mathrm{PGL}_n(q_0) \cap G$, then $|x^G \cap H|$ is roughly $|x^G|^{\frac{1}{2}}$ for all $x \in H$ (see also [18, Example 2.17]).

Proposition 4.2.2 *Let G be an almost simple group with socle $\mathrm{P\Omega}_{2m}^{\varepsilon}(q)$ where $m \geqslant 4$ and $q = p^f$. Let $H \leqslant G$ be a maximal nonsubspace subgroup and let $x \in G$ be nontrivial. Then*

$$\mathrm{fpr}(x, G/H) < \frac{2}{q^{m-2+2/(m+1)}}$$

Moreover,

(i) *if $f \geqslant 2$ and either $\nu(x) \geqslant 2$ or $x \notin \mathrm{PGO}_{2m}^{\varepsilon}(q)$, then*

$$\mathrm{fpr}(x, G/H) < \frac{3}{q^{2m-5+3/m-\ell}}$$

where $\ell = 0$, unless H has type $\mathrm{GL}_m^{\pm}(q)$, in which case $\ell = 2$

(ii) *if* $\mathrm{soc}(G) = \mathrm{P\Omega}_8^+(q)$ *and* H *is almost simple with socle* $G_2(q)$ *or* $\mathrm{PSL}_3^{\pm}(q)$, *then*

$$\mathrm{fpr}(x, G/H) < \frac{2}{q^{9/2}}.$$

Proof We will assume that x has prime order. The first bound is an immediate consequence of [17, Corollary 2], which in turn Burness deduces from Theorem 4.2.1.

Now let us consider part (i). Write $T = \mathrm{P\Omega}_{2m}^{\varepsilon}(q)$. From the bounds presented in [15, Section 3], if $x \in \mathrm{PGO}_{2m}^{\varepsilon}(q)$ and $\nu(x) \geqslant 2$, then

$$|x^G| \geqslant |x^T| \geqslant \frac{2^{\delta_{2,p}}}{8}\left(\frac{q}{q+1}\right)q^{4m-6}$$

and if $x \in \mathrm{Aut}(T) \setminus \mathrm{PGO}_{2m}^{\varepsilon}(q)$, then by [15, Corollary 3.49],

$$|x^G| \geqslant \frac{1}{8}q^{m(m-1/2)} \geqslant \frac{1}{4}\left(\frac{q}{q+1}\right)q^{4m-6}.$$

Theorem 4.2.1 now implies that if $\nu(x) \geqslant 2$ or $x \notin \mathrm{PGO}_{2m}^{\varepsilon}(q)$, then

$$\mathrm{fpr}(x, G/H) < |x^G|^{-\frac{1}{2}+\frac{1}{2m}+\iota} < \frac{\left(8/2^{\delta_{2,p}} \cdot \frac{q+1}{q}\right)^{1/2}}{q^{(4m-6)(\frac{1}{2}-\frac{1}{2m}-\iota)}} \leqslant \frac{3}{q^{2m-5-(4m-6)\iota}},$$

where $\iota = 0$ unless H has type $\mathrm{GL}_m^{\pm}(q)$ and $\iota = (2m-2)^{-1}$, as claimed in (i).

We now turn to part (ii). By [41, Theorem 7.1], if $x \in H \cap \mathrm{PGO}_8^+(q)$, then $\nu(x) \geqslant 3$, so from the bounds in [15, Section 3],

$$|x^G| > \frac{1}{4}\left(\frac{q}{q+1}\right)q^{12}.$$

In addition, by [15, Corollary 3.49], if $x \in \mathrm{Aut}(T) \setminus \mathrm{PGO}_8^+(q)$, then

$$|x^G| > \frac{1}{8}q^{14} \geqslant \frac{1}{4}\left(\frac{q}{q+1}\right)q^{12}.$$

Therefore, by Theorem 4.2.1, we conclude that

$$\mathrm{fpr}(x, G/H) < |x^G|^{-3/8} \leqslant \frac{\left(4 \cdot \frac{q+1}{q}\right)^{3/8}}{q^{9/2}} \leqslant \frac{2}{q^{9/2}}.$$

□

Proposition 4.2.3 *Let G be an almost simple group with socle $\mathrm{PSU}_n(q)$ where $n \geqslant 7$. Let $H \leqslant G$ be a maximal nonsubspace subgroup and let $x \in G$ be nontrivial. Then*

$$\mathrm{fpr}(x, G/H) < \frac{2}{q^{n-3+2/n}}.$$

Proof We may assume that $x \in H$. By [14, Lemma 2.1], $|x^G| > \frac{1}{4}q^{2n-2}$. Therefore, if H does not have type $\mathrm{Sp}_n(q)$, then Theorem 4.2.1 implies that

$$\mathrm{fpr}(x, G/H) < \frac{2}{q^{(2n-2)(1/2-1/n)}} = \frac{2}{q^{n-3+2/n}}.$$

For the remainder of the proof we can assume that H has type $\mathrm{Sp}_n(q)$. By [15, Corollary 3.38], if $x \in \mathrm{PGU}_n(q)$ and $\nu(x) > 1$, then

$$|x^G| > \frac{1}{2}\left(\frac{q}{q+1}\right)q^{4(n-2)},$$

and if $x \in \mathrm{P\Gamma U}_n(q) \setminus \mathrm{PGU}_n(q)$ has odd order, then, by [15, Lemma 3.48]

$$|x^G| > \frac{1}{2}\left(\frac{q}{q+1}\right)q^{2n^2/3-5/3}.$$

In both cases, we obtain the desired bound. Now assume that $x \in \mathrm{PGU}_n(q)$ and $\nu(x) = 1$. Since $x \in H$ we know that $x = [J_2, J_1^{n-2}]$ and we can compute

$$|x^G| > \frac{q^{2n-1}}{2(q+1)} \quad \text{and} \quad |x^G \cap H| \leqslant (2 - \delta_{2,p})|x^H| < q^n,$$

which gives

$$\mathrm{fpr}(x, G/H) < \frac{2(q+1)}{q^{n-1}} < \frac{2}{q^{n-3+2/n}}.$$

Finally assume that $x \in \mathrm{P\Gamma U}_n(q)$ is an involutory graph automorphism. Here the bounds in the statement hold since [14, Lemma 3.48] implies that

$$|x^G| > \frac{1}{2}\left(\frac{q}{q+1}\right)q^{(n^2-n-4)/2}. \qquad \square$$

The rest of this chapter is dedicated to deriving sharp upper bounds on fixed point ratios of nonsubspace actions of low-dimensional almost simple unitary groups.

Theorem 4.2.4 *Let G be an almost simple group with socle $\mathrm{PSU}_n(q)$ where $3 \leqslant n \leqslant 6$. Assume that $q \geqslant 11$ if $n \in \{3, 4\}$. Let $H \leqslant G$ be a maximal nonsubspace subgroup. Let $x \in G$ be nontrivial.*

(i) *If* $n \in \{5, 6\}$, *then*

$$\mathrm{fpr}(x, G/H) \leqslant (q^4 - q^3 + q^2 - q + 1)^{-1}.$$

(ii) *If* $n \in \{3, 4\}$ *and* H *does not have type* $\mathrm{Sp}_4(q)$, *then*

$$\mathrm{fpr}(x, G/H) \leqslant (q^2 - q + 1)^{-1}.$$

(iii) *If* $n = 4$ *and* H *has type* $\mathrm{Sp}_4(q)$, *then*

$$\mathrm{fpr}(x, G/H) \leqslant (2, q + 1) \cdot \frac{q^4 + 1}{q^5 + q^2}.$$

The following will be used in the proof of Theorem 4.2.4 and also in Chap. 6.

Proposition 4.2.5 *Let* $n \geqslant 6$ *and let* G *be an almost simple group with socle* $T = \mathrm{PSL}_n^{\varepsilon}(q)$. *Let* H *be a maximal subgroup of* G *such that* $T \nleqslant H$. *If* $x \in H \cap \mathrm{PGL}_n^{\varepsilon}(q)$ *has prime order and satisfies* $\nu(x) \leqslant 2$, *then one of the following holds*

(i) $H \in \mathcal{C}_1 \cup \mathcal{C}_2 \cup \mathcal{C}_5 \cup \mathcal{C}_8$
(ii) H *appears in Fig. 4.2*
(iii) $n \in \{6, 7\}$, $q = p$, $H \in \mathcal{S}$ *and* $\mathrm{soc}(H)$ *appears in Fig. 4.3.*

	type of H	n	ε	q	x
$\mathcal{C}_3$	$\mathrm{GL}_m(q^2)$	$2m$	$+$	any	$[J_2^2, J_1^{n-4}]$
					$[\lambda, \lambda^q, I_{n-2}]$ with $\lvert\lambda\rvert \mid q^2 - 1$
$\mathcal{C}_4$	$\mathrm{GL}_2^{\varepsilon}(q) \otimes \mathrm{GL}_m^{\varepsilon}(q)$	$2m$	$\pm$	any	$[J_2^2, J_1^{n-4}]$
					$[\lambda I_2, I_{n-2}]$ with $\lvert\lambda\rvert \mid q + 1$
$\mathcal{C}_6$	$2^6.\mathrm{Sp}_6(2)$	8	$\pm$	p	$\nu(x) = 2$
$\mathcal{C}_7$	$\mathrm{GL}_2(q) \wr S_3$	8	$+$	any	$\nu(x) = 2$
$\mathcal{S}$	$\mathrm{PSL}_3^{\varepsilon}(q)$	6	$\pm$	odd	$[-I_2, I_4]$

Fig. 4.2 The subgroups in Proposition 4.2.5(ii)

n	$\mathrm{soc}(H)$	conditions
7	$\mathrm{PSU}_3(3)$	$p \equiv \varepsilon \pmod 3$, $p \geqslant 5$
6	A_6	$p \equiv \varepsilon \pmod 3$, $p \geqslant 5$
	A_7	$p \equiv \varepsilon \pmod 3$, $p \geqslant 5$
	$\mathrm{PSL}_3(4)$	$p \equiv \varepsilon \pmod 3$, $p \geqslant 5$
	$\mathrm{PSU}_4(3)$	$p \equiv \varepsilon \pmod 3$
	M_{12}	$\varepsilon = +$ and $p = 3$
	M_{22}	$\varepsilon = -$ and $p = 2$

Fig. 4.3 The subgroups in Proposition 4.2.5(iii)

Proof Assume neither (i) nor (iii) hold. Then, by [41, Theorem 7.1], we have one of

(a) $H \in \mathcal{C}_3$
(b) $H \in \mathcal{C}_4$
(c) $n = 8, q = p \equiv \varepsilon \pmod 4$ and $H \in \mathcal{C}_6$ has type $2^6.\mathrm{Sp}_6(2)$
(d) $n = 8, \varepsilon = +$ and $H \in \mathcal{C}_7$ has type $\mathrm{GL}_2(q) \wr S_3$
(e) $n = 6, p > 2, H \in \mathcal{S}$ and $\mathrm{soc}(H) = \mathrm{PSL}_3^{\varepsilon}(q)$ is embedded via the symmetric-square of the natural module for $\mathrm{PSL}_3^{\varepsilon}(q)$.

We need to prove that the only cases that (a)–(e) give rise to are those in Fig. 4.2.

For (a), the conclusion is given by [20, Lemma 5.3.2], noting that G does not have any degree two field extension subgroups if $T = \mathrm{PSU}_n(q)$.

We now turn to (b). Assume that $x = x_1 \otimes x_2$ centralises a tensor product decomposition $V = V_1 \otimes V_2$ where $1 < \dim V_1 \leqslant \dim V_2$. Then [55, Lemma 3.7] implies that $n = 2m,\ \nu(x) = 2,\ \dim V_1 = 2,\ \dim V_2 = m,\ \nu(x_1) = 0,\ \nu(x_2) = 1$. Without loss of generality $x_1 = I_2$. If x is unipotent, then $x_2 = [J_2, J_1^{m-2}]$ and $x = [J_2^2, J_1^{n-4}]$. If x is semisimple, then $x_2 = [\lambda, I_{m-1}]$ and $x = [\lambda I_2, I_{n-2}]$ where $\lambda \in \mathbb{F}_{q^2}$ and $|\lambda|$ divides $q + 1$.

For (c), [15, Lemma 6.3] implies that $\nu(x) > 1$.

Now assume that (d) holds. Here x stabilises a tensor product decomposition $V = V_1 \otimes V_2 \otimes V_3$ where $\dim V_i = 2$. From the discussion in (b), x permutes the factors nontrivially. Now [20, Lemma 5.7.2] implies that $\nu(x) > 1$.

Finally consider (e). First assume that x is unipotent. A direct computation verifies that the possible Jordan forms on $\mathbb{F}_p^6$ of order p elements of $\mathrm{GL}_3(p)$ acting on the symmetric square are $[J_3, J_2, J_1]$ and $[J_5, J_1]$ if $p > 3$ or $[J_3^2]$ if $p = 3$, so $x \notin H$. Now assume that x is semisimple. Then the eigenvalues of x are of the form $\alpha^2, \beta^2, \gamma^2, \alpha\beta, \alpha\gamma, \beta\gamma$. Since $\nu(x) \leqslant 2$, at least 4 of these eigenvalues are equal. Therefore, without loss of generality, $\alpha\beta = \alpha\gamma$, so $\beta = \gamma$ and the eigenvalues of x are in fact $\alpha^2, \alpha\beta, \alpha\beta, \beta^2, \beta^2, \beta^2$. Since the eigenvalues of x are not all equal, we know that $\alpha \neq \beta$ and therefore $\alpha\beta \neq \beta^2$. This implies that $\alpha^2 = \beta^2$, so $\beta = -\alpha$ and we conclude that $x = [-\alpha I_2, \alpha I_4] = [-I_2, I_4]$ modulo scalars. □

Before proving Theorem 4.2.4 we handle several cases in a series of lemmas.

Lemma 4.2.6 *Theorem 4.2.4 is true if* $(n, q) \in \{(3, 11), (5, 2), (6, 2)\}$.

Proof This is a straightforward computation in Magma. □

For the rest of this chapter, $3 \leqslant n \leqslant 6$ and $q \geqslant 11$ if $n \in \{3, 4\}$. In addition, G is an almost simple group with socle $\mathrm{PSU}_n(q)$, $H \leqslant G$ is maximal and $x \in G$ has prime order.

Let us specify some particular elements that will demand extra attention.

$$\begin{gathered}
[J_2, J_1^3],\ [J_2^2, J_1],\ [\lambda, I_4],\ [\lambda, \lambda, I_3] \in \mathrm{PGU}_5(q) \\
[J_2, J_1^4],\ [\lambda, I_5] \in \mathrm{PGU}_6(q)
\end{gathered} \tag{4.2}$$

where $\lambda \in \mathbb{F}_{q^2}^{\times}$ and $|\lambda|$ is a prime divisor of $q + 1$.

Notation 4.2.7 Let X be a finite subset of a group G and let r be prime. Then we write

$$I_r(X) = \{x \in X \mid |x| = r\} \qquad i_r(X) = |I_r(X)| \qquad i_{\max}(X) = \max_{r \text{ prime}} i_r(X).$$

As in the proof of Proposition 4.1.3, in the proofs that follow, we will extensively refer to the information presented in [20, Chapter 3] on the conjugacy classes of elements of prime order in almost simple classical groups, but for clarity of exposition we will not constantly cite this source. In particular, conjugacy class sizes that are asserted in these proofs can be deduced from the centraliser orders summarised in [20, Appendix B].

Lemma 4.2.8 *Let H have type $\mathrm{Sp}_n(q)$. Then*

$$\mathrm{fpr}(x, G/H) \leqslant \begin{cases} (2, q+1)(q^4+1)(q^5+q^2)^{-1} & \text{if } n = 4 \\ (q^4 - q^3 + q^2 - q + 1)^{-1} & \text{if } n = 6. \end{cases}$$

Proof Write $H_0 = H \cap \mathrm{PGU}_4(q)$ and $|x| = r$. We will assume that $n = 4$ as the arguments are very similar for $n = 6$, but it is enlightening to note that if $n = 6$ and $x = [J_2, J_1^4]$ then there is a unique class of elements with this Jordan form in $\mathrm{PSU}_6(q)$ and in $\mathrm{PGSp}_6(q)$, so

$$\frac{|x^G \cap H|}{|x^G|} \leqslant \frac{|x^{\mathrm{PGSp}_6(q)}|}{|x^{\mathrm{PSU}_6(q)}|} = \frac{(q^6-1)(q+1)}{(q^6-1)(q^5+1)} = \frac{1}{q^4 - q^3 + q^2 - q + 1}.$$

First assume that $x \in \mathrm{PGU}_4(q)$ and $r \neq p$. Assume that $x \in H$. For now assume that $r > 2$. By [20, Proposition 3.3.1 and Lemma 3.4.3], conjugacy of odd order semisimple elements in $\mathrm{PSp}_4(q)$ and $\mathrm{PGU}_4(q)$ is determined by eigenvalues. Therefore, $x^{\mathrm{PGU}_4(q)} \cap \mathrm{PGSp}_4(q) = x^{\mathrm{PGSp}_4(q)}$, so by a straightforward calculation,

$$\frac{|x^G \cap H|}{|x^G|} = \frac{|x^{\mathrm{PGSp}_4(q)}|}{|x^{\mathrm{PGU}_4(q)}|} \leqslant \frac{q^2+1}{q(q^3+1)}$$

with equality if $x = [\lambda, \lambda, \lambda^{-1}, \lambda^{-1}]$, where $r = |\lambda|$ divides $q - 1$, and this is sufficient to establish the desired bound.

Now assume that $r = 2 \neq p$. If $|x^G \cap H| \neq 0$, then x does not have a one-dimensional 1-eigenspace, so x (as an element of $\mathrm{PGU}_4(q)$) has type t_2 or t_2', so $|x^{\mathrm{PSU}_4(q)}| \geqslant \frac{1}{2}q^4(q^2+1)(q^2-q+1)$. There are four classes of semisimple involutions in $\mathrm{PGSp}_4(q)$, with centralisers of order

$$|C_{\mathrm{PGSp}_4(q)}(t_1)| = 2|\mathrm{Sp}_2(q)|^2, \quad |C_{\mathrm{PGSp}_4(q)}(t_1')| = 2|\mathrm{Sp}_2(q^2)|$$
$$|C_{\mathrm{PGSp}_4(q)}(t_2)| = 2|\mathrm{GL}_2(q)|, \quad |C_{\mathrm{PGSp}_4(q)}(t_2')| = 2|\mathrm{GU}_2(q)|.$$

Therefore,

$$\frac{|x^G \cap H|}{|x^G|} \leqslant \frac{i_2(\mathrm{PGSp}_4(q))}{|x^{\mathrm{PSU}_4(q)}|} \leqslant \frac{2q^4(q^2+2)}{q^4(q^2+1)(q^2-q+1)} < \frac{2(q^4+1)}{q^5+q^2}.$$

Next assume that $x \in \mathrm{PGU}_n(q)$ and $r = p$. Here the possible Jordan forms in H are $[J_2, J_1^2]$, $[J_2^2]$ and (if $p \geqslant 5$) $[J_4]$; let k be 1, 2, $(4, q+1)$ in these three cases, respectively. There is a unique $\mathrm{PGSp}_4(q)$-class of elements of one of these Jordan forms, and there is a unique $\mathrm{PGU}_4(q)$-class, which splits into k distinct $\mathrm{PSU}_4(q)$-classes. This gives

$$\frac{|x^G \cap H|}{|x^G|} \leqslant k \cdot \frac{|x^{\mathrm{PGSp}_4(q)}|}{|x^{\mathrm{PGU}_4(q)}|}.$$

It is easy to compute that

$$\frac{|x^{\mathrm{PGSp}_4(q)}|}{|x^{\mathrm{PGU}_4(q)}|} \leqslant \frac{1}{q^2-q+1}$$

with equality if x has Jordan form $[J_2, J_1^2]$, so the claimed bound holds.

Finally assume that $x \in \mathrm{P\Gamma U}_4(q) \setminus \mathrm{PGU}_4(q)$. If x is a field automorphism of (odd) order r, then

$$\frac{|x^G \cap H|}{|x^G|} \leqslant \frac{|x^{\mathrm{PGSp}_4(q)}|}{|x^{\mathrm{PSU}_4(q)}|} \leqslant \frac{4|\,\mathrm{PGSp}_4(q)||\,\mathrm{PGU}_4(q^{1/r})|}{|\,\mathrm{PGSp}_4(q^{1/r})||\,\mathrm{PGU}_4(q)|} \leqslant \frac{4}{q^{2/3}(q^2-q+1)},$$

which gives the bound.

It remains to assume that x is a graph automorphism. We refer to the graph automorphisms as *symplectic* or *non-symplectic* according to the structure of their centraliser (see, for example, [15, Definition 3.47]). Let γ be a symplectic-type graph automorphism, and write $\widetilde{G} = \mathrm{PGU}_4(q)$ and $\widetilde{H} = C_{\widetilde{G}}(\gamma) = \mathrm{PGSp}_4(q)$. Then $\langle T, x\rangle \leqslant \widetilde{G}{:}\langle\gamma\rangle$ and $\langle H_0, x\rangle \leqslant \widetilde{H} \times \langle\gamma\rangle$; moreover,

$$x^G \cap H \subseteq \{h \in \widetilde{H} \mid h^2 = 1 \text{ and } (h\gamma)^{\widetilde{G}} = x^{\widetilde{G}}\}.$$

The conjugacy classes of involutions in $\widetilde{H}$ are labelled t_1, t_1', t_2, t_2' if $p \neq 2$ and a_2, b_1, c_2 if $p = 2$. From the proof of [15, Proposition 8.1], if $|h| = 2$, then $h\gamma$ is $\widetilde{G}$-conjugate to γ if and only if h has type t_2 or t_2' if $p \neq 2$ or type a_2 if $p = 2$. Therefore, if x is not symplectic-type, then for even q

$$\begin{aligned}\frac{|x^G \cap H|}{|x^G|} &\leqslant \frac{|b_1^{\widetilde{H}}| + |c_2^{\widetilde{H}}|}{|x^{\mathrm{PSU}_4(q)}|} = \frac{|C_{\mathrm{PSp}_4(q)}(b_1)|}{|\,\mathrm{PGU}_4(q)|}\left(\frac{|\,\mathrm{PSp}_4(q)|}{q^4(q^2-1)} + \frac{|\,\mathrm{PSp}_4(q)|}{q^4}\right)\\ &= \frac{1}{q^2(q^3+1)(q^4-1)}\Big((q^4-1) + (q^2-1)(q^4-1)\Big) = \frac{2}{q^3+1},\end{aligned}$$

and for odd $q \geqslant 5$, by [50, Proposition 4.5.5], $|x^G| \geqslant \frac{1}{2}|\operatorname{PGU}_4(q) : \operatorname{PGO}_4^{\varepsilon}(q)|$, so

$$\frac{|x^G \cap H|}{|x^G|} \leqslant \frac{|t_2^{\widetilde{H}}| + |(t_2')^{\widetilde{H}}|}{|x^{\operatorname{PSU}_4(q)}|} \leqslant \frac{|\operatorname{PGO}_4^{\varepsilon}(q)|}{|\operatorname{PGU}_4(q)|}\left(\frac{|\operatorname{PGSp}_4(q)|}{2|\operatorname{GL}_2(q)|} + \frac{|\operatorname{PGSp}_4(q)|}{2|\operatorname{GU}_2(q)|}\right)$$

$$= \frac{4 \cdot (q^2+1)}{(q^3+1)(q^2+\varepsilon)} \leqslant \frac{1}{q^2-q+1}.$$

If x is symplectic-type, then for even q

$$\frac{|x^G \cap H|}{|x^G|} = \frac{1 + |a_2^{\operatorname{Sp}_4(q)}|}{|x^{\operatorname{PSU}_4(q)}|} = \frac{|\operatorname{Sp}_4(q)|}{|\operatorname{PSU}_4(q)|} \cdot \left(1 + \frac{|\operatorname{Sp}_4(q)|}{q^4(q^2-1)}\right) = \frac{q^2}{q^3+1},$$

and for odd q, by [50, Proposition 4.5.6], $|x^G| = \frac{1}{2}|\operatorname{PGU}_4(q) : \operatorname{PGSp}_4(q)|$, so

$$\frac{|x^G \cap H|}{|x^G|} \leqslant \frac{1 + |t_1^{\widetilde{H}}| + |(t_1')^{\widetilde{H}}|}{|x^{\operatorname{PSU}_4(q)}|} = \frac{2|\operatorname{PGSp}_4(q)|}{|\operatorname{PGU}_4(q)|}\left(1 + \frac{|\operatorname{PGSp}_4(q)|}{2|\operatorname{Sp}_2(q)|^2} + \frac{|\operatorname{PGSp}_4(q)|}{2|\operatorname{Sp}_2(q^2)|}\right)$$

$$= \frac{2}{q^2(q^3+1)}\left(1 + \tfrac{1}{2}q^2(q^2+1) + \tfrac{1}{2}q^2(q^2-1)\right) = \frac{2(q^4+1)}{q^2(q^3+1)}.$$

This completes the proof. □

Lemma 4.2.9 *Let H have type $\operatorname{GU}_n(q^{1/k})$ or $\operatorname{SO}_n^{\varepsilon}(q)$. Then*

$$\operatorname{fpr}(x, G/H) \leqslant \begin{cases} (q^2-q+1)^{-1} & \text{if } n \in \{3,4\} \\ (q^4-q^3+q^2-q+1)^{-1} & \text{if } n \in \{5,6\} \text{ and } x \text{ in (4.2).} \end{cases}$$

Proof Write $|x| = r$ and $H_0 = H \cap \operatorname{PGU}_n(q)$. We begin by considering $x \in \operatorname{PGU}_n(q)$. Now $x^G \cap H$ is a subset of all the elements of $\operatorname{Inndiag}(\operatorname{soc}(H_0))$ with the same eigenvalues as x if x is semisimple or the same Jordan form as x if x is unipotent. Using this estimate, together with the information on the conjugacy classes of unitary and orthogonal groups in [20, Chapter 3], it is easy to verify the result. We just give the details when $n = 6$ as the arguments are very similar in the remaining cases.

First assume that $r = p$, so $x = [J_2, J_1^4]$ and $|x^G| \geqslant |x^{\operatorname{PSU}_6(q)}| \geqslant \frac{(q^5+1)(q^6-1)}{6(q+1)}$. If H has type $\operatorname{GU}_6(q^{1/k})$, then $|x^G \cap H| \leqslant |x^{\operatorname{PGU}_6(q^{1/k})}| \leqslant \frac{(q^{5/3}+1)(q^2-1)}{q^{1/3}+1}$, so

$$\frac{|x^G \cap H|}{|x^G|} \leqslant \frac{6(q+1)(q^{5/3}+1)(q^2-1)}{(q^5+1)(q^6-1)(q^{1/3}+1)} < \frac{q+1}{q^5+1} = \frac{1}{q^4-q^3+q^2-q+1}.$$

If H has type $\operatorname{SO}_6^{\varepsilon}(q)$ then p is odd, so $|x^G \cap H| = 0$ as $\operatorname{SO}_n^{\varepsilon}(q)$ does not contain elements with Jordan form $[J_2, J_1^{n-2}]$ in odd characteristic.

Now assume $r \neq p$, so $x = [\lambda, I_5]$ where $|\lambda|$ is 2 or a primitive prime divisor of q^2-1. Here, $|x^{\mathrm{PGU}_6(q)}| = \frac{q^5(q^6-1)}{q+1}$, and $|x^{\mathrm{PGU}_6(q)} \cap \mathrm{PGU}_6(q^{1/k})| \leqslant |x^{\mathrm{PGU}_6(q^{1/k})}| \leqslant \frac{q^{5/3}(q^2-1)}{q^{1/3}+1}$ gives the bound if H has type $\mathrm{GU}_6(q^{1/k})$. Now assume that H has type $\mathrm{SO}_6^{\varepsilon}(q)$. In this case, if $x^G \cap H$ is not empty, then $\lambda = -1$. Here $x^G \cap H \subseteq x_{\square}^H \cup x_{\boxtimes}^H$ where $x_\delta \in \mathrm{PGO}_6^{\varepsilon}(q)$ acts as $-I_1 \perp I_5$ with respect to an orthogonal decomposition $U \perp U^{\perp}$ where U is a nondegenerate 1-space with discriminant δ. Therefore,

$$|x^G \cap H| \leqslant |x_{\square}^{\mathrm{PGO}_6^{\varepsilon}(q)}| + |x_{\boxtimes}^{\mathrm{PGO}_6^{\varepsilon}(q)}| = \frac{1}{2}q^2(q^3+1) + \frac{1}{2}q^2(q^3-1) = q^5$$

and we conclude that

$$\frac{|x^G \cap H|}{|x^G|} \leqslant \frac{q^5(q+1)}{q^5(q^6-1)} < \frac{1}{q^4-q^3+q^2-q+1}.$$

We now turn to the case where $x \in \mathrm{P\Gamma U}_n(q) \setminus \mathrm{PGU}_n(q)$ (so $n \in \{3,4\}$ and $q \geqslant 11$). For now assume that x is a field automorphism. If H has type $\mathrm{GU}_n(q^{1/k})$, then

$$\frac{|x^G \cap H|}{|x^G|} \leqslant \frac{|H||C_G(x)|}{|G|} \leqslant \frac{(n, q+1)|\,\mathrm{PGU}_n(q^{1/k})||\,\mathrm{PGU}_n(q^{1/r})|}{|\,\mathrm{PGU}_n(q)|}$$

which gives the claimed bound since $k, r \geqslant 3$ and $q \geqslant 8$.

Now assume that H has type $\mathrm{SO}_n^{\varepsilon}(q)$. By [20, Proposition 3.5.20],

$$|x^G \cap H| \leqslant i_r(H_0 x) \leqslant |x^{\mathrm{Inndiag}(H_0)}|$$

and this gives the desired bound. For instance, if $n = 3$, then

$$\frac{|x^G \cap H|}{|x^G|} \leqslant \frac{|x^{\mathrm{Inndiag}(H_0)}|}{|x^{\mathrm{PSU}_3(q)}|} \leqslant \frac{(3, q+1)|\,\mathrm{PSO}_3(q)||\,\mathrm{PGU}_3(q^{1/r})|}{|\,\mathrm{PSO}_3(q^{1/r})||\,\mathrm{PGU}_3(q)|}$$

which allows us to conclude that

$$\frac{|x^G \cap H|}{|x^G|} \leqslant \frac{3 \cdot q(q^2-1) \cdot q(q^{2/3}-1)(q+1)}{q^3(q^2-1)(q^3+1) \cdot q^{1/3}(q^{2/3}-1)} < \frac{1}{q^2-q+1}.$$

Finally assume that x is an involutory graph automorphism. We follow the proof of [15, Proposition 5.1] (where the relevant case of Theorem 4.2.1 is proved). First assume that H has type $\mathrm{GU}_n(q^{1/k})$. Then x induces an involutory graph automorphism on H of the same type as it induces on G (that is, symplectic on both or non-symplectic on both). We obtain the desired bound. For example, if $n = 3$ and q is odd, then

$$\frac{|x^G \cap H|}{|x^G|} \leqslant \frac{|x^{\mathrm{Inndiag}(H_0)}|}{|x^{\mathrm{PSU}_3(q)}|} = \frac{(3, q+1)|\,\mathrm{PGU}_3(q^{1/k})||\,\mathrm{PSO}_3(q)|}{|\,\mathrm{PSO}_3(q^{1/l})||\,\mathrm{PGU}_3(q)|} < \frac{1}{q^2-q+1}$$

as we computed above.

Now assume that H has type $\mathrm{SO}_n^{\varepsilon}(q)$, where we follow the proof of [15, Proposition 8.2]. If x is non-symplectic,

$$\frac{|x^G \cap H|}{|x^G|} \leqslant \frac{1+i_2(H_0)}{|x^G|} \leqslant |\,\mathrm{PGO}_n^{\eta}(q)||\,\mathrm{PSU}_n(q)|2(q+1)q^{1+2\delta_{n,4}},$$

which gives the result; in particular, if $n = 3$, then

$$\frac{|x^G \cap H|}{|x^G|} \leqslant \frac{2(3, q+1)}{q(q^2-q+1)} \leqslant \frac{1}{q^2-q+1}.$$

If x is symplectic, then $n = 4$ and from the splitting of $x^G \cap H$ into H-classes described in the proof of [15, Proposition 8.2], we obtain $|x^G \cap H| \leqslant 2q^2$, so

$$\frac{|x^G \cap H|}{|x^G|} \leqslant \frac{2(4, q+1)q^2}{q^2(q^3+1)} = \frac{2(4, q+1)}{(q^2-q+1)(q+1)} \leqslant \frac{1}{q^2-q+1}. \qquad \square$$

Lemma 4.2.10 *Let $n = 3$ and let H have type $\mathrm{GU}_1(q^3)$. Then*

$$\mathrm{fpr}(x, G/H) \leqslant (q^2-q+1)^{-1}.$$

Proof Write $H \cap \mathrm{PGU}_3(q) = H_0 = B{:}\langle\phi\rangle$ where $B \leqslant C_{q^2-q+1}$ and $|\phi| = 3$. Let $x \in H$ with $|x| = r$.

First assume that $x \in \mathrm{PGU}_3(q)$. The order of any prime order element in B is a primitive prime divisor of q^6-1 and any element in $H_0 \setminus B$ has order 3 (for it is conjugate to ϕ or ϕ^2). If r is a primitive prime divisor of q^6-1, then $|x^{\mathrm{PGU}_3(q)}| = q^3(q+1)(q^2-1)$ and $|x^{\mathrm{PGU}_3(q)} \cap H_0| = 3$. Now assume that $r = 3$, so $|x^G \cap H_0| = i_3(H) = 2(q^2-q+1)$. If $p = 3$, then $x = [J_3]$, so $|x^G| \geqslant q(q^2-1)(q^3+1)$. If $p \neq 3$, then $x = [\xi, \xi^{-1}, 1]$ where $|\xi| = 3$, so $|x^G| \geqslant q^3(q-1)(q^2-q+1)$. The required bound holds in every case.

Now assume that $x \in \mathrm{P\Gamma U}_3(q) \setminus \mathrm{PGU}_3(q)$. If $r \geqslant 5$, then

$$\frac{|x^G \cap H|}{|x^G|} \leqslant \frac{|Bx|}{|x^G|} = (q^2-q+1) \cdot \frac{|\,\mathrm{PGU}_3(q^{1/r})|}{|\,\mathrm{PSU}_3(q)|} < (q^2-q+1)^{-1},$$

and if $r = 3$, then $|x^G \cap H| = 0$, since all elements of order 3 in H are contained in $H_0 \leqslant \mathrm{PGU}_3(q)$. Finally if $r = 2$, then x is a graph automorphism and

$$\frac{|x^G \cap H|}{|x^G|} \leqslant \frac{|Bx|}{|x^G|} = (q^2 - q + 1) \cdot \frac{|\mathrm{PSO}_3(q)|}{|\mathrm{PSU}_3(q)|} < (q^2 - q + 1)^{-1}.$$

This completes the proof. □

Lemma 4.2.11 *Let H have type $\mathrm{GU}_{n/k}(q) \wr S_k$ or $\mathrm{GL}_{n/2}(q^2)$. Then*

$$\mathrm{fpr}(x, G/H) \leqslant \begin{cases} (q^2 - q + 1)^{-1} & \text{if } n \in \{3, 4\} \\ (q^4 - q^3 + q^2 - q + 1)^{-1} & \text{if } n \in \{5, 6\} \text{ and } x \text{ in (4.2).} \end{cases}$$

Proof We prove this lemma only when $n = 4$ and H has type $\mathrm{GU}_1(q) \wr S_4$ since the other cases are similar (bearing in mind, only the elements in (4.2) need to be considered when $n \in \{5, 6\}$). Write $H \cap \mathrm{PGU}_4(q) = H_0 = B{:}S_4$, where $B \leqslant C_{q+1}^4/\Delta$ with $\Delta = \{(\lambda, \lambda, \lambda, \lambda) \mid \lambda \in C_{q+1}\}$.

First assume that $x \in \mathrm{PGU}_4(q)$. Let us begin by considering the case where $x \in B$. Then x is diagonal and it is easy to compute $|x^{\mathrm{PGU}_4(q)}|$. In addition, note that $((\lambda_1, \lambda_2, \lambda_3, \lambda_4)\Delta)^G \cap B$ is the set of elements $(\lambda_{1\sigma}, \lambda_{2\sigma}, \lambda_{3\sigma}, \lambda_{4\sigma})$ for some permutation $\sigma \in S_4$. Therefore, one of the following hold, for distinct $\lambda, \mu, \nu \in \mathbb{F}_{q^2}^{\times}$ of order r dividing $q + 1$,

x	$\lvert x^{\mathrm{PGU}_4(q)} \cap B\rvert$
$[\lambda, I_3]$	4
$[\lambda, \lambda, I_2]$	6
$[\lambda, \mu, I_2]$	12
$[\lambda, \mu, \nu, 1]$	24

Now consider the case where $x \in H_0 \setminus B$. Let $\sigma \in S_4$ be the permutation that x induces on the factors of B. Then by [20, Lemma 5.2.6], one of the following holds, where $|\xi| = 3$.

σ^{S_4}	x (if $r = p$)	x (if $r \neq p$)	$\lvert x^{\mathrm{PGU}_4(q)} \cap (H \setminus B)\rvert$
$(1\,2\,3)^{S_4}$	$[J_3, J_1]$	$[\xi, \xi^{-1}, I_2]$	$8(q+1)^2$
$(1\,2)(3\,4)^{S_4}$	$[J_2^2]$	$[-I_2, I_2]$	$3(q+1)$
$(1\,2)^{S_4}$	$[J_2, J_1^2]$	$[-1, I_3]$	$6(q+1)$
		$[-I_2, I_2]$	$6(q+1)$

(Regarding the final two rows of the table above, there are two G-classes of elements that transpose two factors: one negates one of the fixed factors and the other acts trivially on both.) Let us now justify the final column in this table. We concentrate on the case where $\sigma \in (1\,2\,3)^{S_4}$ as the remaining cases are similar. First note that $|\sigma^{S_4}| = 8$. Now assume that x induces $\sigma = (1\,2\,3)$ on the factors of B. It is easy to check that an element $x = (\lambda_1, \lambda_2, \lambda_3, 1)\Delta\sigma \in H_0$ has order 3 if and only if $\lambda_1\lambda_2\lambda_3 = 1$, so there are $(q+1)^2$ choices for x.

With this information, it is easy to check that the required bound holds.

Now assume that $x \in \mathrm{P\Gamma U}_4(q) \setminus \mathrm{PGU}_4(q)$. If x is a field automorphism, then

$$\frac{|x^G \cap H|}{|x^G|} \leqslant \frac{(4, q+1)|H||\,\mathrm{PGU}_4(q^{1/r})|}{|\,\mathrm{PGU}_4(q)|} < \frac{1}{q^2 - q + 1}.$$

Finally assume that x is an involutory graph automorphism. The argument for this case is given in detail in the proof of [16, Proposition 2.7] and the bound can easily be verified. In particular, if x is symplectic, then x acts as a double transposition on the four factors of B, so

$$\frac{|x^G \cap H|}{|x^G|} \leqslant \frac{3(q+1)\cdot(4, q+1)|\,\mathrm{PGSp}_4(q)|}{|\,\mathrm{PGU}_4(q)|} \leqslant \frac{3(4, q+1)}{q^2(q^2 - q + 1)} < \frac{1}{q^2 - q + 1}.$$

This completes the proof. □

We are now ready to prove Theorem 4.2.4.

Proof of Theorem 4.2.4 By Lemma 4.2.6, we may (and we will) assume that $(n, q) \notin \{(3, 11), (5, 2), (6, 2)\}$. In addition, we will assume that (G, H, x) does not appear in Lemmas 4.2.9–4.2.11 as the required bound was shown to hold in these cases.

Case 1: $n = 3$ Consulting [8, Tables 8.5 and 8.6], we see that $H \in \mathcal{C}_6 \cup \mathcal{S}$. The conjugacy classes of prime order elements of $\mathrm{P\Gamma U}_3(q)$, together with the orders of the respective centralisers, are given in [20, Appendix B] and with this information it is easy to check that $|x^G| \geqslant (q^2 - 1)(q^2 - q + 1)$, with equality if $x = J_2 \perp I_1$. Therefore,

$$\frac{|x^G \cap H|}{|x^G|} \leqslant \frac{i_{\max}(H)}{|x^G|} \leqslant \frac{i_{\max}(H)}{(q^2 - 1)(q^2 - q + 1)}.$$

Therefore, it suffices to prove that $i_{\max}(H) \leqslant q^2 - 1$. First assume that $H \in \mathcal{C}_6$. Here $q \equiv 2 \pmod 3$ and H has type $3^{1+2}.Q_8$. It is easy to check that

$$i_{\max}(H) \leqslant |H| \leqslant 216 \leqslant q^2 - 1,$$

since $q \geqslant 17$ in this case. Now assume that $H \in \mathcal{S}$ and write $S = \mathrm{soc}(H)$. If $S = \mathrm{PSL}_2(7)$ and $q \equiv 3, 5, 6 \pmod 7$, then $q \geqslant 13$ and a computation in Magma shows

$$i_{\max}(H) \leqslant i_{\max}(\mathrm{Aut}(S)) = 84 \leqslant q^2 - 1.$$

Similarly, if $S = A_6$ and $q \equiv 11, 14 \pmod{15}$, then $q \geqslant 29$ and

$$i_{\max}(H) \leqslant 360 \leqslant q^2 - 1.$$

Case 2: $n = 4$ By [8, Tables 8.10 and 8.11], $H \in \mathcal{C}_6$ has type $2^4{:}\,\mathrm{Sp}_4(2)$ or $H \in \mathcal{S}$ is an almost simple group with socle $S \in \{\mathrm{PSL}_2(7), A_7, \mathrm{PSU}_4(2)\}$. By Lemma 4.2.6, we assume that $q \geqslant 11$ and, proceeding as in the previous case, we obtain

$$\frac{|x^G \cap H|}{|x^G|} \leqslant \frac{i_{\max}(H)}{|x^G|} < \frac{1}{q^2 - q + 1}.$$

Case 3: $n = 5$ By Theorem 4.2.1, we obtain the desired bound provided that

$$|x^G| \geqslant (q^4 - q^3 + q^2 - q + 1)^{10/3}. \tag{4.3}$$

If $x \in \mathrm{Aut}(\mathrm{PSU}_5(q)) \setminus \mathrm{PSU}_5(q)$, then [15, Lemma 3.48] gives (4.3). From the information in [20, Appendix B], we see that $|x^G| > q^{40/3}$ unless x is conjugate to one of

$$[J_2, J_1^3],\ [J_2^2, J_1],\ [\lambda, I_4],\ [\lambda, \lambda, I_3]. \tag{4.4}$$

Therefore, for the remainder of this case we will assume that x is one of these specific elements.

The maximal subgroups of G are given in [8, Tables 8.20 and 8.21], and we see that either H has type $\mathrm{GU}_1(q^5)$ or $H \in \mathcal{C}_6 \cup \mathcal{S}$. In the first case, $x^G \cap H$ is empty by [20, Lemma 5.3.2]. Now assume that $H \in \mathcal{C}_6 \cup \mathcal{S}$. In this case, it suffices to prove that $|x^G| \geqslant i_{\max}(H)(q^4 - q^3 + q^2 - q + 1)$, whenever $x^G \cap H$ is nonempty.

First assume that $H \in \mathcal{S}$. In this case, $q \geqslant 5$ and $\mathrm{soc}(H)$ is either $\mathrm{PSL}_2(11)$ or $\mathrm{PSU}_4(2)$, so

$$i_{\max}(H) \leqslant i_{\max}(\mathrm{Aut}(S)) = 170 \leqslant q^4 - 1.$$

Observe that $|x^G| \geqslant (q^4 - 1)(q^4 - q^3 + q^2 - q + 1)$, with equality if $x = [J_2, I_3]$, so we obtain the desired bound.

Now assume that $H \in \mathcal{C}_6$. Here either $q = p \equiv 4 \pmod 5$ or $q = p^2$ with $p \equiv 2, 3 \pmod 5$ and H has type $5^{1+2}{:}\,\mathrm{Sp}_2(5)$. In particular,

$$i_{\max}(H) \leqslant 3124 \leqslant q(q^3 + 1)(q^4 - 1).$$

If $|x^G \cap H| \neq 0$, then $\nu(x) \geqslant 2$ by [20, Lemma 5.6.3], so

$$|x^G| \geqslant q(q^3 + 1)(q^4 - 1)(q^4 - q^3 + q^2 - q + 1).$$

Case 4: n $= 6$ By Theorem 4.2.1, it suffices to show that

$$|x^G| \geqslant (q^4 - q^3 + q^2 - q + 1)^3, \tag{4.5}$$

noting that the parameter ι is 0 since H does not have type $\mathrm{Sp}_6(q)$. If $x \notin \mathrm{PGU}_6(q)$, then (4.5) is given by [15, Lemma 3.48]. Now assume that $x \in \mathrm{PGU}_6(q)$. If $\nu(x) \geqslant 2$, then, [15, Corollary 3.38] gives (4.5), so it remains to assume that $\nu(x) = 1$. Proposition 4.2.5 implies that $|x^G \cap H| = 0$ unless H has type $\mathrm{GL}_3(q^2)$ or $H \in \mathcal{S}$ and

$$\mathrm{soc}(H) \in \{A_6,\ A_7,\ \mathrm{PSL}_3(4),\ \mathrm{PSU}_4(3)\}. \tag{4.6}$$

First assume that H has type $\mathrm{GL}_3(q^2)$. We claim that $|x^G \cap H| = 0$. Write $V = \mathbb{F}_{q^2}^6$. Then H stabilises a decomposition $V = U \oplus U^*$ where U is a maximal totally singular subspace of V and let $B \leqslant H$ centralise this decomposition. By [20, Lemma 5.2.6], $|x^G \cap (H \setminus B)| = 0$, and all of the elements of B are of the form $g \oplus g^{-(q)\mathsf{T}}$, so $|x^G \cap B| = 0$ also.

Now assume that $H \in \mathcal{S}$. Here we see that $|x^G| \geqslant q^5(q^2 - q + 1)(q^3 - 1)$ and, via computation in Magma, $i_{\max}(H) \leqslant (q^2-q+1)(q^3-1)$, which gives the result in the familiar way. This completes the proof. □

Chapter 5
Orthogonal Groups

5.1 Introduction

We now turn our focus to proving our main results on uniform spread: Theorems 4 and 5. The aim of this chapter is to prove Theorems 4 and 5 for even-dimensional orthogonal groups. We will consider the remaining unitary groups in Chap. 6. For this entire chapter, write $q = p^f$ and

$$\mathcal{T} = \{\mathrm{P\Omega}^{\varepsilon}_{2m}(q) \mid m \geqslant 4 \text{ and } \varepsilon \in \{+,-\}\} \tag{5.1}$$

$$\mathcal{A} = \{\langle T, \theta \rangle \mid T \in \mathcal{T} \text{ and } \theta \in \mathrm{Aut}(T)\}. \tag{5.2}$$

The main results of this chapter are the following.

Theorem 5.1 *If $G \in \mathcal{A}$, then $u(G) \geqslant 2$.*

Theorem 5.2 *Let (G_i) be a sequence of groups in $\mathcal{A}$ with $\mathrm{soc}(G_i) = \mathrm{P\Omega}^{\varepsilon_i}_{2m_i}(q_i)$. Then $u(G_i) \to \infty$ if $q_i \to \infty$.*

Let us now discuss the proofs. Let $G = \langle T, \theta \rangle \in \mathcal{A}$ with $T \in \mathcal{T}$. As we explained in the introduction, to prove that $u(G) \geqslant k$ for some $k \geqslant 1$, we adopt the probabilistic approach introduced by Guralnick and Kantor in [38] (see Sect. 2.1). Recall that this approach has three stages. First we must fix an element $s \in G$. In order for s^G to witness $u(G) \geqslant k$, the element s cannot be contained in a proper normal subgroup of G, so we may assume that $s \in T\theta$. Consequently we need to understand the conjugacy classes in the coset $T\theta$. We then study the set $\mathcal{M}(G, s)$ of maximal subgroups of G that contain s, before showing that every prime order element $x \in G$ satisfies

$$P(x, s) \leqslant \sum_{H \in \mathcal{M}(G,s)} \mathrm{fpr}(x, G/H) < \frac{1}{k}.$$

S. Harper, *The Spread of Almost Simple Classical Groups*, Lecture Notes in Mathematics 2286, https://doi.org/10.1007/978-3-030-74100-6_5

We must first determine the automorphisms θ it suffices to consider, and this will require a detailed analysis of the automorphism group of T.

Generically, θ will be a field or graph-field automorphism (possibly multiplied by a nontrivial element of Inndiag(T)). In this case, we view $G = \langle T, \theta \rangle$ from the perspective of algebraic groups, which allows us to employ Shintani descent. The main idea, therefore, is to write $\mathrm{Inndiag}(T) = X_{\sigma^e}$ and $\theta \in \mathrm{Inndiag}(T)\sigma$ for a suitable connected algebraic group X, Steinberg endomorphism σ and integer $e > 1$ (see Example 3.1.5). We may then select an element $s \in T\theta$ as the preimage under the Shintani map of a judiciously chosen element $x \in X_\sigma$ (see Proposition 5.4.7).

However, unlike in the previous study of symplectic and odd-dimensional orthogonal groups in [43], it will not always be possible to write $\mathrm{Inndiag}(T) = X_{\sigma^e}$ and $\theta \in \mathrm{Inndiag}(T)\sigma$ for the same Steinberg endomorphism σ, and we need to apply Shintani descent differently and use Lemma 3.4.1 (see Examples 3.1.5 and 3.4.2).

Of course, there are other types of automorphisms θ that must be considered. If θ is diagonal, then we can employ methods similar to those used by Breuer, Guralnick and Kantor in [11]. When θ is an involutory graph automorphism (for example, a reflection), then we must necessarily select an element $s \in T\theta$ that fixes a 1-space of $\mathbb{F}_q^{2m}$, which makes bounding $P(x, s)$ more difficult (recall from Chap. 4 that the fixed point ratio of an element of prime order on 1-spaces can be as large as roughly q^{-1}). Consequently, we give a constructive proof that some specific pairs of elements generate G in addition to a probabilistic argument which deals with the general case (see Proposition 5.5.7). This constructive argument is of a different flavour to much of the rest of the proofs. Finally, when $T = \mathrm{P\Omega}_8^+(q)$, we must also take into account triality graph and graph-field automorphisms.

In light of the above discussion, it is natural to partition our analysis into the following cases

(I) $\theta \in \mathrm{P\Gamma O}_{2m}^{\varepsilon}(q) \setminus \mathrm{PGO}_{2m}^{\varepsilon}(q)$
(II) $\theta \in \mathrm{PGO}_{2m}^{\varepsilon}(q)$
(III) $\theta \in \mathrm{Aut}(\mathrm{P\Omega}_8^+(q)) \setminus \mathrm{P\Gamma O}_8^{\varepsilon}(q)$.

In Cases I and II, we define the following two subcases

(a) $G \cap \mathrm{PGO}_{2m}^{\varepsilon}(q) \leqslant \mathrm{PDO}_{2m}^{\varepsilon}(q)$
(b) $G \cap \mathrm{PGO}_{2m}^{\varepsilon}(q) \not\leqslant \mathrm{PDO}_{2m}^{\varepsilon}(q)$.

Recall that the group $\mathrm{PDO}_{2m}^{\varepsilon}(q)$ (defined in (2.3) and (2.4) in Sect. 2.2) is nothing other than $\mathrm{Inndiag}(\mathrm{P\Omega}_{2m}^{\varepsilon}(q))$.

In short, Cases I(b) and II(b) are more difficult than Cases I(a) and II(a). Case I(b) is exactly the situation in which Shintani descent does not apply directly, and in Case II(b) we encounter the obstacle of graph automorphisms we discussed above. We will partition Case III further but we postpone the details of this until the introduction to Sect. 5.6.

This chapter is organised as follows. We begin with two sections that determine general properties about almost simple orthogonal groups. In particular, in Sect. 5.2, we will determine the conjugacy classes of the outer automorphism group and

Sect. 5.3 will introduce the elements that will play a central role in our proofs. We will then prove Theorems 5.1 and 5.2, considering Cases I–III in Sects. 5.4–5.6, respectively.

5.2 Automorphisms

Let $T \in \mathcal{T}$. The main result of this section is Proposition 5.2.15, which details the automorphisms $\theta \in \mathrm{Aut}(T)$ it suffices to consider to prove Theorems 5.1 and 5.2.

5.2.1 Preliminaries

Let us fix some notation. For $g \in \mathrm{Aut}(T)$, write $\ddot{g}$ for the set Tg. Therefore, $\mathrm{Out}(T) = \{\ddot{g} \mid g \in \mathrm{Aut}(T)\}$. We begin with a preliminary elementary observation, which we will also use in Sect. 6.2.

Lemma 5.2.1 *Let $S = \langle a \rangle{:}\langle b \rangle$ be a semidirect product of finite cyclic groups. For all $i > 0$ there exist $j, k \in \mathbb{N}$ such that $\langle ab^i \rangle = \langle a^j b^k \rangle$ and k divides $|b|$.*

Proof Let $i > 0$. We repeatedly use the fact that, since $\langle a \rangle \trianglelefteq S$, for all $l \in \mathbb{N}$

$$(ab^i)^l \in \langle a \rangle b^{il}. \tag{5.3}$$

Write $|b| = n$, and let k divide n and satisfy $\langle b^i \rangle = \langle b^k \rangle$. Now let r be the least positive integer such that $b^{ir} = b^k$. By (5.3), $|ab^i| = s|b^i|$. Let d be the product of the distinct prime divisors of s which do not divide r. Then, by (5.3), $(ab^i)^{r+d|b^i|} = a^j b^k$ for some $j \in \mathbb{N}$. Therefore, $\langle a^j b^k \rangle \leqslant \langle ab^i \rangle$.

Recall that $|ab^i| = s|b^i|$. Note that $(r + d|b^i|, |b^i|) = (r, |b^i|) = 1$ as $\langle b^{ir} \rangle = \langle b^i \rangle$. Let t be a prime divisor of s. If t does not divide r, then t does not divide $r + d|b^i|$ since t divides d. Now assume that t divides r. If t divides $r + d|b^i|$, then t divides $d|b^i|$, so t divides $|b^i|$ since t does not divide d. However, this implies that t divides $(r, |b^i|) = 1$, which is a contradiction. Therefore, t does not divide $r + d|b^i|$. Consequently, $(r + d|b^i|, s) = 1$. We now conclude that $(r + d|b^i|, s|b^i|) = 1$, so $\langle a^j b^k \rangle = \langle ab^i \rangle$, which proves the claim. □

For the remainder of this section, write $n = 2m$, $q = p^f$ and $V = \mathbb{F}_q^n$. Further, let $\mathcal{B}^\varepsilon$ be the basis from (2.7) or (2.8). Write $\mathbb{F}_q^\times = \langle \alpha \rangle$. In addition, if q is odd, then let $\beta \in \mathbb{F}_q^\times$ with $|\beta| = (q-1)_2$ and note that $\alpha, \beta \notin (\mathbb{F}_q^\times)^2 = \{x^2 \mid x \in \mathbb{F}_q^\times\}$.

5.2.2 Plus-Type

Let $T = \mathrm{P}\Omega^+_{2m}(q)$ with $m \geqslant 4$. Fix the standard Frobenius endomorphism $\varphi = \varphi_{\mathcal{B}^+} : (a_{ij}) \mapsto (a_{ij}^p)$ and the standard reflection $r \in \mathrm{PO}^+_{2m}(q)$ from Definition 2.5.1. It will be useful to fix $r_\square$ and $r_\boxtimes$ as the images in $\mathrm{PO}^+_{2m}(q)$ of reflections in vectors of square and nonsquare norm respectively (evidently, if q is even, then we do not use the notation $r_\boxtimes$). In [50, Chapter 2], the symbols $r_\square$ and $r_\boxtimes$ (and also δ, introduced below) refer to elements of $\mathrm{GO}^+_{2m}(q)$, but we prefer to use these symbols for elements of $\mathrm{PGO}^+_{2m}(q)$.

Definition 5.2.2 Let q be odd. With respect to the basis $\mathcal{B}^+$ for $\mathbb{F}_q^{2m}$, define $\hat{\delta}^+ \in \mathrm{GL}_{2m}(q)$ as $\beta I_m \oplus I_m$, which centralises the decomposition $\langle e_1, \ldots, e_m \rangle \oplus \langle f_1, \ldots, f_m \rangle$ and let $\delta^+ \in \mathrm{PGL}_{2m}(q)$ be the image of $\hat{\delta}^+$.

Remark 5.2.3 We comment on Definition 5.2.2.

(i) Note that $\hat{\delta}^+$ is a similarity with $\tau(\hat{\delta}^+) = \beta$ and $\det(\hat{\delta}^+) = \beta^m$.
(ii) We will refer to δ^+ simply as δ if the sign is understood. (A different element $\delta^- \in \mathrm{PDO}^-_{2m}(q)$ will be introduced in Sect. 5.2.3.)
(iii) Our definition of δ differs from that in [50]: both versions centralise the decomposition $\langle e_1, \ldots, e_m \rangle \oplus \langle f_1, \ldots, f_m \rangle$, but we work with $\beta I_m \oplus I_m$ rather than $\alpha I_m \oplus I_m$. However, both versions give the same element $\ddot{\delta}$. To see this, write $k = ((q-1)_{2'} - 1)/2$ and note that

$$(\alpha I_m \oplus I_m) \cdot (\alpha^k I_m \oplus \alpha^{-k} I_m) \cdot \alpha^k I_{2m} = \beta I_m \oplus I_m$$

where $(\alpha^k I_m \oplus \alpha^{-k} I_m) \in \mathrm{GO}^+_{2m}(q)$ and $\alpha^k I_{2m}$ is a scalar.

By [50, Proposition 2.7.3], if $T = \mathrm{P}\Omega^+_{2m}(q)$ with $m \geqslant 5$, then

$$\mathrm{Out}(T) = \begin{cases} \langle \ddot{r}_\square \rangle \times \langle \ddot{\varphi} \rangle \cong C_2 \times C_f & \text{if } q \text{ is even} \\ \langle \ddot{\delta} \rangle \times \langle \ddot{r}_\square \rangle \times \langle \ddot{\varphi} \rangle \cong C_2 \times C_2 \times C_f & \text{if } q \text{ is odd } \& \ D(Q) = \boxtimes \\ \langle \ddot{\delta}, \ddot{r}_\square, \ddot{r}_\boxtimes, \ddot{\varphi} \rangle \cong D_8 \times C_f & \text{if } q \text{ is odd } \& \ D(Q) = \square. \end{cases} \tag{5.4}$$

Now assume that $m = 4$. The group $\mathrm{P}\Omega^+_8(q)$ has a *triality* automorphism τ such that $C_G(\tau) \cong G_2(q)$ (see [27, pp. 200–202]). From [47, Section 1.4], if $T = \mathrm{P}\Omega^+_8(q)$, then

$$\mathrm{Out}(T) = \begin{cases} \langle \ddot{r}_\square, \ddot{\tau} \rangle \times \langle \ddot{\varphi} \rangle \cong S_3 \times C_f & \text{if } q \text{ is even} \\ \langle \ddot{\delta}, \ddot{r}_\square, \ddot{\tau} \rangle \times \langle \ddot{\varphi} \rangle \cong S_4 \times C_f & \text{if } q \text{ is odd.} \end{cases} \tag{5.5}$$

Remark 5.2.4 Let $T = \mathrm{P}\Omega^+_{2m}(q)$. Assume that q is odd and $D(Q) = \square$. By [50, Proposition 2.7.3(iii)], $\langle \ddot{r}_\square, \ddot{r}_\boxtimes, \ddot{\delta} \rangle \cong D_8$. Moreover, if m is even, then

$$|\ddot{r}_\square \ddot{\delta}| = 4, \quad |\ddot{\delta}| = 2, \quad (\ddot{r}_\square \ddot{\delta})^{\ddot{\delta}} = (\ddot{r}_\square \ddot{\delta})^{-1}, \quad (\ddot{r}_\square \ddot{\delta})^2 = \ddot{r}_\square \ddot{r}_\boxtimes,$$

and if m is odd, then

$$|\ddot{\delta}| = 4, \quad |\ddot{r}_\square \ddot{\delta}| = 2, \quad \ddot{\delta}^{\ddot{r}_\square \ddot{\delta}} = \ddot{\delta}^{-1}, \quad \ddot{\delta}^2 = \ddot{r}_\square \ddot{r}_\boxtimes.$$

In both cases, $Z(\langle \ddot{r}_\square, \ddot{r}_\boxtimes, \ddot{\delta} \rangle) = \langle \ddot{r}_\square \ddot{r}_\boxtimes \rangle$.

It will be convenient to write $\mathrm{Out}_0(T)$ for $\mathrm{P\Gamma O}^+_{2m}(q)/T$, so $\mathrm{Out}_0(T) = \mathrm{Out}(T)$ if $m \geqslant 5$ and $|\mathrm{Out}(T) : \mathrm{Out}_0(T)| = 3$ if $m = 4$. Since φ arises from an automorphism of $\mathrm{GL}_{2m}(q)$, the group $\mathrm{Out}_0(T)$ splits as the semidirect product $\langle \ddot{r}_\square, \ddot{r}_\boxtimes, \ddot{\delta} \rangle{:}\langle \ddot{\varphi} \rangle$. If $\ddot{\varphi} \in Z(\mathrm{Out}_0(T))$, then evidently we have $\mathrm{Out}_0(T) \cong D_8 \times C_f$. However, $\ddot{\varphi}$ need not be central in $\mathrm{Out}_0(T)$. In particular, by [50, Proposition 2.7.3(iii)],

$$[\ddot{r}_\square, \ddot{\varphi}] = [\ddot{r}_\boxtimes, \ddot{\varphi}] = 1$$

but

$$\ddot{\varphi} \notin Z(\mathrm{Out}_0(T)) \iff [\ddot{\delta}, \ddot{\varphi}] \neq 1 \iff m \text{ is odd and } p \equiv 3 \pmod 4.$$

If $\ddot{\varphi} \notin Z(\mathrm{Out}_0(T))$, then $\ddot{\delta}$ has order 4 and $\ddot{\delta}^{\ddot{\varphi}} = \ddot{\delta}^{-1}$, which implies that $\mathrm{Out}_0(T) = \langle \ddot{r}_\square, \ddot{r}_\boxtimes, \ddot{\delta} \rangle \times \langle \ddot{r}_\square \ddot{\varphi} \rangle$. In this case, $p \equiv 3 \pmod 4$ and $q \equiv 1 \pmod 4$, so f is even and $\ddot{r}_\square \ddot{\varphi}$ has order f; this shows that $\mathrm{Out}_0(T) \cong D_8 \times C_f$ in this case also.

Remark 5.2.5 Let $m = 4$. In this case $\ddot{\varphi} \in Z(\mathrm{Out}(T))$, and $\{1, \ddot{r}_\square, \ddot{\tau}\}$ is a set of conjugacy class representatives of $\langle \ddot{r}_\square, \ddot{\tau} \rangle \cong S_3$ if q is even and $\{1, \ddot{r}_\square, \ddot{\delta}, \ddot{\delta}\ddot{r}_\square, \ddot{\tau}\}$ is a set of conjugacy class representatives of $\langle \ddot{r}_\square, \ddot{r}_\boxtimes, \ddot{\delta}, \ddot{\tau} \rangle$ if q is odd.

The following lemma provides further information when q is odd and $D(Q) = \square$. It is useful to record the following set of conditions

$$m \text{ is odd } \textit{and } p \equiv 3 \pmod 4 \textit{ and } i \text{ is odd } \textit{and } f \text{ is even.} \tag{5.6}$$

Lemma 5.2.6 *Let $T = \mathrm{P}\Omega^+_{2m}(q)$. Assume that q is odd and $D(Q) = \square$. For $0 \leqslant i < f$, the following hold*

(i) $\ddot{\delta}\ddot{\varphi}^i$ *and* $\ddot{r}_\square \ddot{r}_\boxtimes \ddot{\delta} \ddot{\varphi}^i$ *are* $\mathrm{Out}(T)$*-conjugate*
(ii) $\ddot{\delta}\ddot{r}_\square \ddot{\varphi}^i$ *and* $\ddot{\delta}\ddot{r}_\boxtimes \ddot{\varphi}^i$ *are* $\mathrm{Out}(T)$*-conjugate*
(iii) $\ddot{\varphi}^i$ *and* $\ddot{r}_\square \ddot{r}_\boxtimes \ddot{\varphi}^i$ *are* $\mathrm{Out}(T)$*-conjugate if* (5.6) *holds*
(iv) $\ddot{r}_\square \ddot{\varphi}^i$ *and* $\ddot{r}_\boxtimes \ddot{\varphi}^i$ *are* $\mathrm{Out}(T)$*-conjugate if* (5.6) *does not hold.*

Proof Write $A = \langle \ddot{r}_\square, \ddot{r}_\boxtimes, \ddot{\delta} \rangle$. The description of $\mathrm{Out}_0(T)$ in Remark 5.2.4 allows us to deduce that the conjugacy classes of A are

$$\{\ddot{1}\}, \ \{\ddot{r}_\square \ddot{r}_\boxtimes\}, \ \{\ddot{r}_\square, \ddot{r}_\boxtimes\}, \ \{\ddot{\delta}, \ddot{r}_\square \ddot{r}_\boxtimes \ddot{\delta}\}, \ \{\ddot{\delta}\ddot{r}_\square, \ddot{\delta}\ddot{r}_\boxtimes\}.$$

If the condition (5.6) is not satisfied, then $\ddot{\varphi}^i \in Z(\mathrm{Out}_0(T))$ and (i), (ii) and (iv) follow. Now assume that condition (5.6) is satisfied. In this case $\ddot{r}_\square \ddot{\varphi}^i \in Z(\mathrm{Out}_0(T))$. Writing

$$\begin{aligned} \ddot{\delta}\ddot{\varphi}^i &= \ddot{r}_\boxtimes \ddot{\delta}(\ddot{r}_\square \ddot{\varphi}^i) \quad \text{and} \quad \ddot{r}_\square \ddot{r}_\boxtimes \ddot{\delta}\ddot{\varphi}^i = \ddot{r}_\square \ddot{\delta}(\ddot{r}_\square \ddot{\varphi}^i) \\ \ddot{\delta}\ddot{r}_\square \ddot{\varphi}^i &= \ddot{\delta}(\ddot{r}_\square \ddot{\varphi}^i) \quad \text{and} \quad \ddot{\delta}\ddot{r}_\boxtimes \ddot{\varphi}^i = \ddot{r}_\square \ddot{r}_\boxtimes \ddot{\delta}(\ddot{r}_\square \ddot{\varphi}^i) \\ \ddot{\varphi}^i &= \ddot{r}_\square(\ddot{r}_\square \ddot{\varphi}^i) \quad \text{and} \quad \ddot{r}_\square \ddot{r}_\boxtimes \ddot{\varphi}^i = \ddot{r}_\boxtimes(\ddot{r}_\square \ddot{\varphi}^i) \end{aligned}$$

reveals that (i), (ii) and (iii) hold. □

Recall the definition of $\mathrm{PDO}^+_{2m}(q)$ from Sect. 2.2 (see (2.3) and (2.4)). The following is [50, Proposition 2.7.4], but it can be quickly deduced from (2.12).

Lemma 5.2.7 *Let $T = \mathrm{P\Omega}^+_{2m}(q)$ with $m \geqslant 4$. Then*

$$\mathrm{Inndiag}(T) = \mathrm{PDO}^+_{2m}(q) = \begin{cases} T & \text{if } q \text{ is even} \\ \langle T, \delta \rangle & \text{if } q \text{ is odd and } D(Q) = \boxtimes \\ \langle T, r_\square r_\boxtimes, \delta \rangle & \text{if } q \text{ is odd and } D(Q) = \square. \end{cases}$$

5.2.3 *Minus-Type*

Now let $T = \mathrm{P\Omega}^-_{2m}(q)$ with $m \geqslant 4$. To describe $\mathrm{Out}(T)$ in this case we deviate from [50] and work more in the spirit of [37]. This is because we want to work with a copy of $\mathrm{P\Omega}^-_{2m}(q)$ that arises naturally from the perspective of algebraic groups. However, we do want to be able to concretely work with the action of $\mathrm{P\Omega}^-_{2m}(q)$ on the natural module $\mathbb{F}_q^{2m}$, so we will recover some of the key results from [50, Section 2.8] in our context. In this section, the isomorphism Ψ from Lemma 2.5.4 will be the key tool for relating our two viewpoints.

Recall the standard Frobenius endomorphism $\varphi = \varphi_{\mathcal{B}^+}\colon (a_{ij}) \mapsto (a_{ij}^p)$ and the reflection $r \in \mathrm{PO}^+_{2m}(q)$ from Definition 2.5.1. Recall from Lemma 2.5.4 that $\mathrm{PDO}^-_n(q) = X_{r\varphi^f}\Psi$, where $X = \mathrm{PDO}_n(\overline{\mathbb{F}}_p)$. Define $\psi\colon X\Psi \to X\Psi$ as

$$\psi = \Psi \circ \varphi \circ \Psi^{-1}. \tag{5.7}$$

Then

$$\mathrm{Aut}(T) = \mathrm{PDO}_n^-(q){:}\langle\psi\rangle = \mathrm{P\Gamma O}_n^-(q)$$

and $\psi^f = r\Psi = r$. We use $r_\square$ and $r_\boxtimes$ as in plus-type, but we often, instead, work with the reflection r, which we may assume is contained in $\{r_\square, r_\boxtimes\}$.

If q is odd, then we define a further element.

Definition 5.2.8 Let q be odd. With respect to $\mathcal{B}^+$, define $\Delta \in \mathrm{GO}_{2m}^+(q^2)$ as $\beta I_{m-1} \oplus I_{m-1} \perp [\beta_2, \beta_2^q]$, centralising $\langle e_1, \dots, e_{m-1}\rangle \oplus \langle f_1, \dots, f_{m-1}\rangle \perp \langle e_m, f_m\rangle$, where $\beta_2 \in \mathbb{F}_{q^2}^\times$ has order $(q^2-1)_2$. Let $\hat{\delta}^-$ be $\Delta\Psi$ and $\delta^- \in \mathrm{PGO}_{2m}^-(q)$ its image.

Remark 5.2.9 We comment on Definition 5.2.8.

(i) If the sign $-$ is understood, then we omit reference to it.
(ii) Since $\Delta \in \mathrm{GO}_{2m}^+(q^2)$ is fixed by $r\varphi^f$, we have $\hat{\delta} \in \mathrm{GO}_{2m}^-(q)$.
(iii) Evidently, $\det(\Delta) = \beta^m$, so $\det(\hat{\delta}) = \det(\Delta\Psi) = \beta^m$.
(iv) It is straightforward to verify that $\tau(\Delta) = \beta_2^{q+1} = \beta$, with respect to the plus-type form defined in terms of $\mathcal{B}^+$. This implies that $\tau(\hat{\delta}) = \beta$ with respect to the minus-type form defined in terms of $\mathcal{B}^-$.

Lemma 5.2.10 *Let $T = \mathrm{P\Omega}_{2m}^-(q)$. Then*

$$\mathrm{Inndiag}(T) = \mathrm{PDO}_{2m}^-(q) = \begin{cases} T & \text{if } q \text{ is even} \\ \langle T, \delta\rangle & \text{if } q \text{ is odd.} \end{cases}$$

Proof By (2.12), $\mathrm{Inndiag}(T) = \mathrm{PDO}_{2m}^-(q)$. If q is even, then $\mathrm{PDO}_{2m}^-(q) = T$ (see (2.4)). Now assume that q is odd. Note that $\tau(\hat{\delta}) = \beta$, so $\delta \notin \mathrm{PO}_{2m}^-(q)$. Since $|\,\mathrm{PGO}_{2m}^-(q) : \mathrm{PO}_{2m}^-(q)| = 2$, we deduce that $\mathrm{PGO}_{2m}^-(q) = \langle \mathrm{PO}_{2m}^-(q), \delta\rangle$. Now $\mathrm{PDO}_{2m}^-(q) \cap \mathrm{PO}_{2m}^-(q) = \mathrm{PSO}_{2m}^-(q)$ and $\delta^- \in \mathrm{PDO}_{2m}^-(q)$ since $\det(\hat{\delta}) = \beta^m = \tau(\hat{\delta})^m$, so $\mathrm{PDO}_{2m}^-(q) = \langle \mathrm{PSO}_{2m}^-(q), \delta\rangle$. Since $\mathrm{Inndiag}(T)/T$ is cyclic (see [37, Theorem 2.5.12]) in fact, $\mathrm{PDO}_{2m}^-(q) = \langle T, \delta\rangle$, which completes the proof. □

Remark 5.2.11 In light of Lemma 5.2.10, let us compare our notation for $\mathrm{PGO}_{2m}^-(q)$ with the notation of Kleidman and Liebeck in [50, Section 2.8]. Their symbol $\ddot{r}_\square$ is also our $\ddot{r}_\square$, but their $\ddot{\delta}$ is our $\ddot{r}^m\ddot{\delta}$. Therefore, we may conclude from [50, Section 2.8] that, in our notation, if $D(Q) = \boxtimes$ then $|\ddot{\delta}| = 2$, and if $D(Q) = \square$ then $|\ddot{\delta}| = 4$ with $\ddot{\delta}^2 = \ddot{r}_\square\ddot{r}_\boxtimes$.

We now describe $\mathrm{Out}(T)$.

Lemma 5.2.12 *Let $T = \mathrm{P\Omega}_{2m}^-(q)$. Then*

$$\mathrm{Out}(T) = \begin{cases} \langle\ddot{\psi}\rangle \cong C_{2f} & \text{if } q \text{ is even} \\ \langle\ddot{\delta}\rangle \times \langle\ddot{\psi}\rangle \cong C_2 \times C_{2f} & \text{if } q \text{ is odd and } D(Q) = \boxtimes \\ \langle\ddot{\delta}\rangle{:}\langle\ddot{\psi}\rangle \cong C_4{:}C_{2f} & \text{if } q \text{ is odd and } D(Q) = \square. \end{cases}$$

Proof Recall that $\mathrm{Aut}(T) = \mathrm{Inndiag}(T){:}\langle\psi\rangle$. By Lemma 5.2.10 we see that $\mathrm{Out}(T) = \langle\ddot{\psi}\rangle$ when q is even and $\mathrm{Out}(T) = \langle\ddot{\delta}\rangle{:}\langle\ddot{\psi}\rangle$ when q is odd. Since $|\ddot{\psi}| = |\psi| = 2f$, we have proved the claim when q is even.

Now assume that q is odd. If $D(Q) = \boxtimes$, then, by Remark 5.2.11, $|\ddot{\delta}| = 2$, so $\ddot{\psi}$ centralises $\ddot{\delta}$. It remains to assume that $D(Q) = \square$. In this case, f is necessarily odd (see (2.2)), so $\langle\ddot{\psi}\rangle = \langle\ddot{r}\ddot{\psi}^2\rangle$, since $\psi^f = r$. By Remark 5.2.11, $|\ddot{\delta}| = 4$, so $\ddot{\psi}^2$, having odd order, centralises $\ddot{\delta}$. Since $r_v^{\delta} = r_{v\delta}$, for all nonzero $v \in V$, we know that $\ddot{r}_{\square}^{\ddot{\delta}} = \ddot{r}_{\boxtimes}$. Therefore,

$$\ddot{\delta}^{\ddot{\psi}} = \ddot{\delta}^{\ddot{r}} = \ddot{\delta}\ddot{r}_{\square}\ddot{r}_{\boxtimes} = \ddot{\delta}^{-1}.$$

This completes the proof. □

Remark 5.2.13 Let $T = \mathrm{P}\Omega^-_{2m}(q)$. Assume that q is odd and $D(Q) = \square$. From the proof of Lemma 5.2.12, $|\ddot{\delta}| = 4$, $|\ddot{r}_{\square}| = 2$ and $\ddot{\delta}^{\ddot{r}_{\square}} = \ddot{\delta}^{-1}$, so $\langle\ddot{\delta}, \ddot{r}\rangle \cong D_8$. Moreover, $[\ddot{\delta}, \ddot{\psi}^2] = 1$, so

$$\mathrm{Out}(T) \cong \langle\ddot{\delta}, \ddot{r}\rangle \times \langle\ddot{\psi}^2\rangle \cong D_8 \times C_f.$$

Lemma 5.2.14 *Let $T = \mathrm{P}\Omega^-_{2m}(q)$. Assume that q is odd and $D(Q) = \square$. For $0 \leqslant i < 2f$, the following hold*

(i) *$\ddot{\delta}\ddot{\psi}^i$ and $\ddot{\delta}^{-1}\ddot{\psi}^i$ are* Out(T)*-conjugate*
(ii) *if i is odd, then $\ddot{\psi}^i$ and $\ddot{r}_{\square}\ddot{r}_{\boxtimes}\ddot{\psi}^i$ are* Out(T)*-conjugate.*

Proof From Remark 5.2.13, $\ddot{\delta}^{\ddot{r}} = \ddot{\delta}^{-1}$ and $[\ddot{r}, \ddot{\psi}] = 1$, so $(\ddot{\delta}\ddot{\psi}^i)^{\ddot{r}} = \ddot{\delta}^{-1}\ddot{\psi}^i$. Moreover, if i is odd, then $(\ddot{\psi}^i)^{\ddot{\delta}} = \ddot{\delta}^{-1}\ddot{\delta}^{\ddot{\psi}^i}\ddot{\psi}^i = \ddot{\delta}^{-1}\ddot{\delta}^{-1}\ddot{\psi}^i = \ddot{r}_{\square}\ddot{r}_{\boxtimes}\ddot{\psi}^i$. □

5.2.4 Conjugacy of Outer Automorphisms

For this section, define

$$d = \begin{cases} 1 \text{ if } \varepsilon = + \\ 2 \text{ if } \varepsilon = -. \end{cases} \tag{5.8}$$

Proposition 5.2.15 *Let $G \in \mathcal{A}$ with* soc(G) $= T = \mathrm{P}\Omega^{\varepsilon}_{2m}(q)$*. Then G is* Aut(T)*-conjugate to $\langle T, \theta\rangle$ for exactly one of the following*

(i) *θ in Row (1) of Fig. 5.1*
(ii) *θ in Row (2) of Fig. 5.1, if q is odd*
(iii) *θ in Row (3) of Fig. 5.1, if q is odd and $D(Q) = \square$*
(iv) *$\theta = \tau\varphi^i$ where i is 0 or a proper divisor of f, if $m = 4$ and $\varepsilon = +$.*

Before proving Proposition 5.2.15 we must comment on Fig. 5.1.

	I(i)	I(ii)	I(iii)	I(iv)	I(v)	II(a)	II(b)	
ε	$+$	$+$	$+$	$-$	$-$			
	φ^i	$r\varphi^i$	$r\varphi^i$	ψ^i	ψ^i	1	r	(1)
θ	$\delta\varphi^i$	$\delta r\varphi^i$	$\delta r\varphi^i$	$\delta\psi^i$	$\delta\psi^i$	δ	δr	(2)
	$r_\square r_\boxtimes \varphi^i$	$r_\square r_\boxtimes r\varphi^i$		$r_\square r_\boxtimes \psi^i$		$r_\square r_\boxtimes$		(3)
df/i	any	even	odd	odd	even			
notes	★	†						

Note: i is a proper divisor of df and the notes are given in Remark 5.2.16

Fig. 5.1 The relevant automorphisms θ when $T = \mathrm{P}\Omega^{\varepsilon}_{2m}(q)$

Remark 5.2.16 Let us explain how to read Fig. 5.1.

(i) In Case I, the possibilities for θ depend on whether ε is $+$ or $-$. Moreover, we have used the conditions on ε and i to define five subcases. Observe that Case I(a) is the union of Cases I(i), I(ii) and I(iv), whereas Case I(b) is the union of Cases I(iii) and I(v). We will often refer to these subcases.

(ii) In Case II, the description is uniform for both signs ε, but we have noted which of Cases II(a) and II(b) the automorphism θ arises in.

(iii) Observe that part (iv) of Proposition 5.2.15 corresponds to Case III. We define subcases of Case III in the introduction to Sect. 5.6.

(iv) We now comment on the notes.

★ $\ddot{\varphi}^i$ and $\ddot{r}_\square \ddot{r}_\boxtimes \ddot{\varphi}^i$ are $\mathrm{Out}(T)$-conjugate if the condition (5.6) holds.
† $\ddot{r}\ddot{\varphi}^i$ and $\ddot{r}_\square \ddot{r}_\boxtimes \ddot{r}\ddot{\varphi}^i$ are $\mathrm{Out}(T)$-conjugate *unless* (5.6) holds.

We now give the proof of Proposition 5.2.15.

Proof Write $G = \langle T, g \rangle$ where $g \in \mathrm{Aut}(T)$. We will study the $\mathrm{Out}(T)$-conjugacy classes, since two groups $\langle T, g_1 \rangle$ and $\langle T, g_2 \rangle$ are $\mathrm{Aut}(T)$-conjugate if and only if $\ddot{g}_1$ and $\ddot{g}_2$ are $\mathrm{Out}(T)$-conjugate.

Begin by assuming that $\varepsilon = +$. By inspecting the structure of $\mathrm{Out}(T)$ given above, it is manifest that we may write $g = h\varphi^i$ where h is a product of diagonal and graph automorphisms. Assume for now that $i > 0$. Since $\langle \ddot{h}, \ddot{\varphi} \rangle = \langle \ddot{h} \rangle{:}\langle \ddot{\varphi} \rangle$, by Lemma 5.2.1, there exist $j, k \in \mathbb{N}$ with k dividing f such that $\langle \ddot{h}\ddot{\varphi}^i \rangle = \langle \ddot{h}^j \ddot{\varphi}^k \rangle$. Therefore, we assume that i divides f. That is, we may assume that $\ddot{g} = \ddot{h}\ddot{\varphi}^i$ where h is a product of diagonal and graph automorphisms and where either $i = 0$ or i divides f. If either $m \geqslant 5$ and q is even, or q is odd and $D(Q) = \boxtimes$, then $\ddot{g}$ is clearly equal to an automorphism in Fig. 5.1. Moreover, if $m = 4$ or if q is odd and $D(Q) = \square$, then Remark 5.2.5 and Lemma 5.2.6 establish that $\ddot{g}$ is $\mathrm{Out}(T)$-conjugate to an automorphism featuring in the statement of the proposition. This proves the result when $\varepsilon = +$.

Now assume that $\varepsilon = -$. As in plus-type, we can assume that $\ddot{g} = \ddot{h}\ddot{\psi}^i$ where h is a diagonal automorphism and where either $i = 0$ or i divides $2f$. Noting that $\psi^f = r$, it follows that $\ddot{g}$ is $\mathrm{Out}(T)$-conjugate to an automorphism $\ddot{\theta}$ in the statement, where we apply Lemma 5.2.14 when q is odd and $D(Q) = \square$. This completes the proof. $\square$

Remark 5.2.17 Proposition 5.2.15 determines the Out(T)-classes if $T = \mathrm{P}\Omega^{\varepsilon}_{2m}(q)$ and Proposition 6.2.6 does when $T = \mathrm{PSL}^{\varepsilon}_{n}(q)$. For completeness let us record these classes when T is $\mathrm{PSp}_{2m}(q)$ or $\Omega_{2m+1}(q)$ (see [50, Propositions 2.4.4 and 2.6.3]). In this case, if $G = \langle T, g \rangle$ for $g \in \mathrm{Aut}(T)$, then G is Aut(T)-conjugate to $\langle T, \theta \rangle$ for exactly one of the automorphisms θ below, where i is a divisor of f:

T	conditions	Out(T)	
$\mathrm{PSp}_{2m}(q)$	$p = 2$ & $m > 2$	$\langle \ddot{\varphi} \rangle \cong C_f$	φ^i
	$p = 2$ & $m = 2$	$\langle \ddot{\rho} \rangle \cong C_{2f}$	ρ^j (for $j \mid 2f$)
	$p > 2$	$\langle \ddot{\delta}, \ddot{\varphi} \rangle \cong C_2 \times C_f$	φ^i, $\delta\varphi^i$
$\Omega_{2m+1}(q)$	$p > 2$	$\langle \ddot{r}_{\square}\ddot{r}_{\boxtimes}, \ddot{\varphi} \rangle \cong C_2 \times C_f$	φ^i, $r_{\square} r_{\boxtimes} \varphi^i$

Remark 5.2.18 We note in passing that our approach of considering each simple group T and each automorphism $\theta \in \mathrm{Aut}(T)$ (with the reductions justified by Proposition 5.2.15) allows us to avoid mentioning the classical groups that Bray, Holt and Roney-Dougal [7] highlight are not well-defined (such as the one often referred to as $\mathrm{P\Sigma O}^{+}_{2m}(q)$).

Now that we have established the cases to consider, let us conclude this section by immediately handling some small orthogonal groups. This result can be established by way of computation in MAGMA (see Sect. 2.7).

Proposition 5.2.19 *Let $G \in \mathcal{A}$. Then $u(G) \geqslant 2$ if the socle of G is one of the following groups*

$$\Omega^{\pm}_{8}(2),\ \mathrm{P}\Omega^{\pm}_{8}(3),\ \Omega^{\pm}_{8}(4),\ \Omega^{\pm}_{10}(2),\ \Omega^{\pm}_{12}(2). \tag{5.9}$$

5.3 Elements

For this section, write $V = \mathbb{F}_q^n$ where $n \geqslant 1$ and $q = p^f$. Write $\mathbb{F}_q^{\times} = \langle \alpha \rangle$. We will define several *types* of semisimple elements in symplectic and orthogonal groups that will play an important part in the proofs, both later in this chapter and in Chap. 6. (Indeed our reason for considering symplectic groups is that, in addition to orthogonal groups, they arise as centralisers of graph automorphisms in unitary groups and these feature significantly in Sect. 6.5.2.)

5.3.1 Preliminaries

The following technical result will be useful.

Lemma 5.3.1 *Let r divide $q^n - 1$ but not divide $q^k - 1$ for $k < n$. Assume that $g \in \mathrm{GL}_n(q)$ has an eigenvalue over $\overline{\mathbb{F}}_p$ of order r. Then g is irreducible on $\mathbb{F}_q^n$ and the eigenvalues of g over $\overline{\mathbb{F}}_p$ are $\lambda, \lambda^q, \ldots, \lambda^{q^{n-1}}$, which are all distinct.*

Proof Let $\lambda \in \overline{\mathbb{F}}_p$ be an eigenvalue of g of order r and let ϕ be the minimal polynomial of λ over $\mathbb{F}_q$. Since r does not divide $q^k - 1$ for $k < n$, the element λ is contained in $\mathbb{F}_{q^n}$ and is not contained in any proper subfield of $\mathbb{F}_{q^n}$. Therefore the degree of ϕ is n, so ϕ is the characteristic polynomial of g. This implies that g has an irreducible characteristic polynomial, so, by Lemma 2.3.5, g is irreducible on $\mathbb{F}_q^n$. Moreover, the eigenvalues of g are the roots of ϕ, which are the n distinct Galois conjugates $\lambda, \lambda^q, \ldots, \lambda^{q^{n-1}}$. This completes the proof. □

Applying Lemma 5.3.1 gives the following familiar result.

Lemma 5.3.2 *Let r divide $q^n - 1$ but not divide $q^k - 1$ for $k < n$. Let $\lambda \in \mathbb{F}_{q^n}^\times$ of order r. Then $\mathrm{GL}_n(q)$ has an irreducible element with eigenvalues $\lambda, \lambda^q, \ldots, \lambda^{q^{n-1}}$.*

Proof Consider the field extension embedding $\pi\colon \mathrm{GL}_1(q^n) \to \mathrm{GL}_n(q)$. Now $g = \pi((\lambda)) \in G$ has order r and λ is an eigenvalue of g. Therefore, by Lemma 5.3.1, g is irreducible and has eigenvalues $\lambda, \lambda^q, \ldots, \lambda^{q^{n-1}}$. □

For the remainder of this section write $n = 2m$. Extending the argument in the previous proof to symplectic and orthogonal groups yields the following two results. We only prove the latter since the former is similar but easier.

Lemma 5.3.3 *Let G be $\mathrm{Sp}_{2m}(q)$ or $\mathrm{SO}^-_{2m}(q)$. Let r divide $q^m + 1$ but not divide $q^k - 1$ for $k < 2m$. Let $\lambda \in \mathbb{F}_{q^{2m}}^\times$ have order r. Then G contains an irreducible element with eigenvalues $\lambda, \lambda^q, \ldots, \lambda^{q^{2m-1}}$.*

Lemma 5.3.4 *Let q be odd and let G be either $\mathrm{GSp}_{2m}(q)$ or $\mathrm{DO}^-_{2m}(q)$. Let r be a divisor of $q^m + 1$ that is divisible by $(q^m + 1)_2$. Assume that $r/2$ does not divide $q^k - 1$ for $k < 2m$. Then G contains an element g of order $(q - 1)r$ such that $\tau(g) = \alpha$ and g^{q-1} is irreducible.*

Proof First assume that $G = \mathrm{GSp}_{2m}(q)$. Let $\lambda \in \mathbb{F}_{q^{2m}}^\times$ have order $(q - 1)r$. The order of λ^{q^m+1} is $(q - 1)r/(q^m + 1, (q - 1)r)$. Since r divides $q^m + 1$,

$$(q^m + 1,\ (q - 1)r) = r\left(\tfrac{1}{r}(q^m + 1),\ q - 1\right) = r,$$

since $(q^m + 1, q - 1) = 2$ and $(q^m + 1)_2$ divides r. Therefore, λ^{q^m+1} has order $q - 1$. Consequently, we may choose λ such that $\lambda^{q^m+1} = \alpha$.

There is a field extension embedding $\pi_1 \colon H \to \mathrm{GSp}_{2m}(q)$, where

$$H = \{h \in \mathrm{GSp}_2(q^m) \mid \tau(h) \in \mathbb{F}_q\} = \{h \in \mathrm{GL}_2(q^m) \mid \det(h) \in \mathbb{F}_q\},$$

where the second equality holds since $\mathrm{GSp}_2(q^m) = \mathrm{GL}_2(q^m)$ and $\tau(h) = \det(h)$ for all $h \in \mathrm{GSp}_2(q^m)$ (see [50, Lemma 2.4.5], for example). In addition, there is a field extension embedding $\pi_2 \colon K \to H$, where

$$K = \{(\mu) \in \mathrm{GL}_1(q^{2m}) \mid \mu^{q^m+1} \in \mathbb{F}_q\}.$$

Now $g = \pi_1(\pi_2((\lambda))) \in G$ has order $(q-1)r$. Moreover,

$$\tau(g) = \tau(\pi_2((\lambda))) = \det(\pi_2((\lambda))) = \lambda^{q^m+1} = \alpha.$$

Now λ is an eigenvalue of g, so λ^{q-1} is an eigenvalue of g^{q-1}. Since λ^{q-1} has order r, by Lemma 5.3.1, g^{q-1} is irreducible.

Now assume that $G = \mathrm{DO}^-_{2m}(q)$. In this case, let $\lambda \in \mathbb{F}^\times_{q^{2m}}$ have order r. There is a field extension embedding $\pi \colon H \to \mathrm{DO}^-_{2m}(q)$, where

$$H = \{h \in \mathrm{DO}^-_2(q^m) \mid \tau(h) \in \mathbb{F}_q\} \cong C_{(q^m+1)(q-1)}.$$

Now fix $h \in \mathrm{DO}^-_2(q^m)$ of order $(q-1)r$ and $\tau(h) = \alpha$. Without loss of generality, the eigenvalues of h are λ and $\alpha\lambda^{-1}$. Let $g = \pi(h)$. Then g has order $(q-1)r$ and $\tau(g) = \tau(h) = \alpha$. Moreover, λ^{q-1} is an eigenvalue of g^{q-1} of order $r/(r, q-1) = r/2$, so Lemma 5.3.1 implies that g^{q-1} is irreducible. This completes the proof. □

Let (G, C) be $(\mathrm{Sp}_{2m}(q), \mathrm{GSp}_{2m}(q))$ or $(\mathrm{O}^+_{2m}(q), \mathrm{GO}^+_{2m}(q))$ and let $V = \mathbb{F}_q^{2m}$ be the natural module for G. Then V admits a decomposition $\mathcal{D}(V)$

$$V = V_1 \oplus V_2 \quad \text{where} \quad V_1 = \langle e_1, \dots, e_m \rangle \text{ and } V_2 = \langle f_1, \dots, f_m \rangle, \tag{5.10}$$

noting that V_1 and V_2 are totally singular m-spaces (with respect to the bases in (2.5) and (2.7)). The following describes the centraliser of the decomposition $\mathcal{D}(V)$.

Lemma 5.3.5 *Let* (G, C) *be* $(\mathrm{Sp}_{2m}(q), \mathrm{GSp}_{2m}(q))$ *or* $(\mathrm{O}^+_{2m}(q), \mathrm{GO}^+_{2m}(q))$. *Then*

(i) $G_{(\mathcal{D}(V))} = \{g \oplus g^{-\mathsf{T}} \mid g \in \mathrm{GL}_m(q)\}$
(ii) $C_{(\mathcal{D}(V))} = \{\lambda g \oplus g^{-\mathsf{T}} \mid g \in \mathrm{GL}_m(q)$ *and* $\lambda \in \mathbb{F}_q^\times\}$
(iii) *If* $g \in \mathrm{GL}_m(q)$ *and* $\lambda \in \mathbb{F}_q^\times$, *then* $\tau(\lambda g \oplus g^{-\mathsf{T}}) = \lambda$.

Proof The matrix of the underlying bilinear form with respect to the basis $(e_1, \dots, e_m, f_1, \dots, f_m)$ is

$$M = \begin{pmatrix} 0 & I_m \\ \pm I_m & 0 \end{pmatrix}.$$

Let $x = g \oplus h \in \mathrm{GL}(V)$ centralise $\mathcal{D}(V)$. If x is a similarity of the form, then, for some $\lambda \in \mathbb{F}_q^\times$, we have $xMx^{-\mathsf{T}} = \lambda M$ and consequently $g = \lambda h^{-\mathsf{T}}$. It is straightforward to see that all such elements are indeed similarities. This proves (ii). Now let $\lambda \in \mathbb{F}_q^\times$ and $g \in \mathrm{GL}(V)$. Write $x = \lambda g \oplus g^{-\mathsf{T}}$. Then $xMx^{-\mathsf{T}} = \lambda M$, so $\tau(x) = \lambda$. This proves (iii) and consequently (i). □

Lemma 5.3.6 *Let G be $\mathrm{Sp}_{2m}(q)$ or $\mathrm{SO}^+_{2m}(q)$. Let r divide $q^m - 1$ but not divide $q^k - 1$ for $k < m$. Then G contains an element of order r that centralises $\mathcal{D}(V)$ and acts irreducibly on both V_1 and V_2.*

Proof By Lemma 5.3.2, there exists an irreducible element $g \in \mathrm{GL}_m(q)$ of order r. The corresponding element $g \oplus g^{-\mathsf{T}} \in G_{(\mathcal{D}(V))}$ satisfies the statement. □

5.3.2 *Types of Semisimple Elements*

Write $V = \mathbb{F}_q^{2m}$ and $\mathbb{F}_q^\times = \langle \alpha \rangle$. By applying the results of Sect. 5.3.1, in this section we will define some important types of semisimple elements in symplectic and orthogonal groups. The general idea that motivates these definitions is that we are interested in elements that stabilise few subspaces, which are contained in particular cosets of $\mathrm{Sp}_{2m}(q)$ in $\mathrm{GSp}_{2m}(q)$ or $\Omega^\pm_{2m}(q)$ in $\mathrm{GO}^\pm_{2m}(q)$ and whose orders are constrained.

We say that a positive integer a is *Mersenne* if $a = 2^b - 1$ for some b. Recall from Sect. 2.6 that $\Phi^*_e(q)$ is the largest divisor of $q^e - 1$ that is coprime to $q^k - 1$ for all $k < e$, and that $\Phi^*_e(q) > 1$ unless $q = 2$ and $e = 6$ or q is Mersenne and $e = 2$.

Definition 5.3.7 Let m be odd and let G be $\mathrm{Sp}_{2m}(q)$ or $\mathrm{SO}^+_{2m}(q)$. An element $g \in G$ has *type* $(2m)^+_q$ if $|g| = \Phi^*_m(q)$ and g centralises $V = V_1 \oplus V_2$ where V_1 and V_2 are totally singular nonisomorphic irreducible $\mathbb{F}_q\langle g \rangle$-modules.

Lemma 5.3.8 *Let G be $\mathrm{Sp}_{2m}(q)$ or $\mathrm{SO}^+_{2m}(q)$ and assume that m is odd. Then G contains an element of type $(2m)^+_q$.*

Proof Theorem 2.6.2 implies that $\Phi^*_e(q) > 1$ and Lemma 5.3.6 establishes that G contains $g \oplus g^{-\mathsf{T}}$ of order r that centralises $\mathcal{D}(V)$ and acts irreducibly on both V_1 and V_2. Since m is odd, the eigenvalue sets of g and $g^{-\mathsf{T}}$ are distinct, so V_1 and V_2 are nonisomorphic. Therefore, $g \oplus g^{-\mathsf{T}}$ has type $(2m)^+_q$. □

Definition 5.3.9 Let G be $\mathrm{Sp}_{2m}(q)$ or $\mathrm{SO}^-_{2m}(q)$. An element $g \in G$ has *type* $(2m)^-_q$ if g is irreducible on V and

$$|g| = \begin{cases} 9 & \text{if } q = 2 \text{ and } m = 3 \\ q+1 & \text{if } q \text{ is Mersenne and } m = 1 \\ \Phi^*_{2m}(q) & \text{otherwise.} \end{cases}$$

Lemma 5.3.10 *Let G be $\mathrm{Sp}_{2m}(q)$ or $\mathrm{SO}^{-}_{2m}(q)$. Then G contains an element of type $(2m)^{-}_{q}$.*

Proof If q is Mersenne and $m = 1$, or $q = 2$ and $m = 3$, then let $r = q^m + 1$; otherwise, let $r = \Phi^*_{2m}(q)$, noting that $r > 1$, by Theorem 2.6.2. Now Lemma 5.3.3 implies that G contains an irreducible element of order r. □

Lemma 5.3.11 *Let $g \in \mathrm{SO}^{\varepsilon}_{2m}(q)$ have type $(2m)^{\varepsilon}_{q}$. Then $g \notin \Omega^{\varepsilon}_{2m}(q)$ if and only if $\varepsilon = -$, $m = 1$ and q is Mersenne.*

Proof First assume that $\varepsilon = -$, $m = 1$ and q is Mersenne. Then $|g| = q + 1$ and $|\Omega^{-}_{2}(q)| = \frac{1}{2}(q + 1)$, so $g \notin \Omega^{-}_{2}(q)$. Now assume otherwise. Therefore, g has odd order, so $g \in \Omega^{\varepsilon}_{2m}(q)$. □

Lemma 5.3.12 *Let g be an element of $\mathrm{Sp}_{2m}(q)$ or $\mathrm{SO}^{\varepsilon}_{2m}(q)$ of type $(2m)^{\varepsilon}_{q}$. Then the eigenvalues of g (over $\overline{\mathbb{F}}_p$) are distinct.*

Proof If $\varepsilon = -$, then g is irreducible, so the characteristic polynomial of g over $\mathbb{F}_q$ is irreducible and the eigenvalues of g are distinct. Now assume that m is odd and $\varepsilon = +$. Then $g = x \oplus x^{-\mathsf{T}}$, centralising the decomposition $\mathcal{D}(V)$ (see (5.10)) where x and $x^{-\mathsf{T}}$ act irreducibly on V_1 and V_2. Therefore, the characteristic polynomial of x is irreducible. Moreover, V_1 and V_2 are nonisomorphic $\mathbb{F}_q\langle x\rangle$-modules, so the characteristic polynomials of x and $x^{-\mathsf{T}}$ are distinct irreducible polynomials. Consequently, g has distinct eigenvalues in this case too. □

Now assume that q is odd. Fix $\beta \in \mathbb{F}_q^{\times}$ with $|\beta| = (q-1)_2$. We will define some variants on the types of elements defined above, which have a very similar action on the natural module. Consequently, in the first instance the reader is encouraged to think of elements of type $(2m)^{\pm}_{q}$ upon encountering ${}^{\Delta}(2m)^{\pm}_{q}$ and ${}^{\Sigma}(2m)^{\pm}_{q}$.

Definition 5.3.13 Let q be odd, let $\varepsilon \in \{+, -\}$ and let G be $\mathrm{GSp}_{2m}(q)$ or $\mathrm{DO}^{\varepsilon}_{2m}(q)$. An element $g \in G$ has *type* ${}^{\Delta}(2m)^{\varepsilon}_{q}$ if $\tau(g) = \beta$ and g^k has type $(2m)^{\varepsilon}_{q}$ where

$$k = \begin{cases} (q^m + 1)_2(q-1)_2 & \text{if } \varepsilon = - \text{ and either } m > 1 \text{ or } q \text{ is not Mersenne} \\ (q-1)_2 & \text{otherwise.} \end{cases}$$

Lemma 5.3.14 *Let q be odd, let $\varepsilon \in \{+, -\}$ and let G be $\mathrm{GSp}_{2m}(q)$ or $\mathrm{DO}^{\varepsilon}_{2m}(q)$.*

(i) *If $\varepsilon = +$ and $m > 1$ is odd, then G contains an element of type ${}^{\Delta}(2m)^{+}_{q}$.*
(ii) *If $\varepsilon = -$, then G contains an element of type ${}^{\Delta}(2m)^{-}_{q}$*

Proof First assume that $\varepsilon = +$. By Lemma 5.3.8, G contains an element $g \oplus g^{-\mathsf{T}}$ of type $(2m)^{+}_{q}$. Let $h = \beta g \oplus g^{-\mathsf{T}}$, noting that $h \in G$ (see Lemma 5.3.5(ii)). We claim that h has type ${}^{\Delta}(2m)^{+}_{q}$. By Lemma 5.3.5(iii), $\tau(h) = \beta$. Now $|g|$ is odd, since $|g| = \Phi^*_m(q)$, and $|\beta| = (q-1)_2$, so $h^{(q-1)_2} = g^{(q-1)_2} \oplus (g^{(q-1)_2})^{-\mathsf{T}}$ has order $|g|$. Therefore, $h^{(q-1)_2}$ has type $(2m)^{+}_{q}$ and, consequently, h has type ${}^{\Delta}(2m)^{+}_{q}$.

Now assume that $\varepsilon = -$. For now assume further that $m > 1$ or q is not Mersenne. Theorem 2.6.2 implies that $r = \Phi^*_{2m}(q) > 1$. By Lemma 5.3.4, there exists an element $g \in G$ of order $r(q^m+1)_2(q-1)$ such that $\tau(g) = \alpha$ and g^{q-1} is irreducible. Let $h = g^{(q-1)_{2'}}$. Then $h^{(q^m+1)_2(q-1)_2}$ has type $(2m)_q^-$ and $\tau(h)$ has order $(q-1)_2$, so without loss of generality, $\tau(h) = \beta$. Therefore, h has type ${}^{\Delta}(2m)_q^-$.

It remains to consider the case where $\varepsilon = -$, $m = 1$ and q is Mersenne. Then Lemma 5.3.4 implies that there exists $g \in G$ of order $q^2 - 1$ such that $\tau(g) = \alpha$ and g^{q-1} is irreducible. As before, $g^{(q-1)_{2'}}$ has type ${}^{\Delta}(2)_q^-$. □

Definition 5.3.15 Let q be odd. An element $g \in \mathrm{SO}^{\varepsilon}_{2m}(q) \setminus \Omega^{\varepsilon}_{2m}(q)$ has *type* ${}^{\Sigma}(2m)_q^{\varepsilon}$ if g^k has type $(2m)_q^{\varepsilon}$ where $k = (q^m - \varepsilon)_2$.

Lemma 5.3.16 *Let q be odd.*

(i) *If $m > 1$ is odd, then $\mathrm{SO}^+_{2m}(q)$ contains an element of type ${}^{\Sigma}(2m)_q^+$.*
(ii) *If $m > 1$, then $\mathrm{SO}^-_{2m}(q)$ contains an element of type ${}^{\Sigma}(2m)_q^-$.*

Proof First assume that $\varepsilon = +$ and $m > 1$ is odd. By Theorem 2.6.2, $r = \Phi^*_m(q) > 1$. Let $\lambda \in \mathbb{F}^{\times}_{q^{2m}}$ have order $r(q^m-1)_2$. By Lemma 5.3.2, $\mathrm{GL}_m(q)$ contains an element of $r(q^m-1)$ and determinant $\lambda^{q^{m-1}+\cdots+q+1}$. Let $h = g \oplus g^{-\mathsf{T}}$. By Lemma 5.3.5(i), $h \in \mathrm{SO}^+_{2m}(q)$. We know that $\lambda \notin (\mathbb{F}^{\times}_{q^m})^2$ since $(q^m-1)_2$ divides the order of λ. Therefore, $\det(g) = \lambda^{q^{m-1}+\cdots+q+1} \notin (\mathbb{F}_q^{\times})^2$. Consequently, $h \notin \Omega^+_{2m}(q)$ by [50, Lemma 4.1.9]. Now $h^{(q^m-1)_2}$ has type $(2m)_q^+$, so h has type ${}^{\Sigma}(2m)_q^+$.

Now assume that $\varepsilon = -$ and $m > 1$. By Theorem 2.6.2, $r = \Phi^*_{2m}(q) > 1$. By Lemma 5.3.3, $\mathrm{SO}^-_{2m}(q)$ contains an irreducible element h of order $r(q^m+1)_2$. By [25, Theorem 4], $(q^m+1)_2$ does not divide the order of a maximal torus of $\Omega^-_{2m}(q)$, so $g \notin \Omega^-_{2m}(q)$. Since $h^{(q^m+1)_2}$ has type $(2m)_q^-$, h has type ${}^{\Sigma}(2m)_q^-$, which completes the proof. □

For all of the elements introduced in this section, if the field size q is clear from the context, then we omit the subscript of q from the notation. However, in general, the field size is pertinent, as Lemma 5.3.17 demonstrates.

Lemma 5.3.17 *Let $m > 1$ and $q = q_0^e$. Let G be $\mathrm{Sp}_{2m}(q)$ or $\mathrm{SO}^{\eta}_{2m}(q)$. Let $g \in G$ have odd order and type $(2m)^{\eta}_{q_0}$. Assume that m is odd if $\eta = +$ and that $(q_0, m) \neq (2, 6)$ if $\eta = -$. Let $t = (m, e)$ and $\varepsilon = \eta^{e/t}$. Then g is similar to $g_1 \oplus \cdots \oplus g_t$ where each g_i has the same order as g and*

(i) *if $\varepsilon = -$, acts irreducibly on a (minus-type) $\frac{2m}{t}$-space*
(ii) *if $\varepsilon = +$, centralises the decomposition of a (plus-type) $\frac{2m}{t}$-space into a dual pair of nonisomorphic totally singular $\frac{m}{t}$-spaces acting irreducibly on both.*

Proof First assume that $\varepsilon = +$. Then $|g| = \Phi^*_m(q_0)$ and the eigenvalue set of g is $\Lambda \cup \Lambda^{-1}$ where $\Lambda = \{\lambda, \lambda^{q_0}, \ldots, \lambda^{q_0^{m-1}}\}$. There are $t = (m, e)$ distinct $\mu \mapsto \mu^q$ orbits on Λ, say $\Lambda_1, \ldots, \Lambda_t$, each of size m/t. Fix $1 \leqslant j \leqslant t$ and $\lambda_j \in \Lambda_j$. By Lemma 5.3.1, there exists an irreducible element $x_j \in \mathrm{GL}_{m/t}(q)$ with eigenvalue set Λ_j. Then $g_j = x_j \oplus x_j^{-\mathsf{T}}$ has eigenvalue set $\Lambda_j \cup \Lambda_j^{-1}$. Therefore, g has the same eigenvalues as $g_1 \oplus \cdots \oplus g_t$. Noting that g is a semisimple element of odd order, Lemma 2.3.7 implies that g is similar to $g_1 \oplus \cdots \oplus g_t$. This proves the claim in this case.

Now assume that $\varepsilon = -$. Then $|g| = \Phi^*_{2m}(q_0)$ and $\Lambda = \{\lambda, \lambda^{q_0}, \ldots, \lambda^{q_0^{2m-1}}\}$ is the eigenvalue set of g. There are $k = (2m, e)$ distinct $\mu \mapsto \mu^q$ orbits of Λ, say $\Lambda_1, \ldots, \Lambda_k$, each of size $2m/k$. Assume for now that $2m/k$ is odd. Then $k = (2m, e) = 2(m, e) = 2t$ and we may assume that $\Lambda_{t+j} = \Lambda_j^{-1}$ for each $1 \leqslant j \leqslant t$. As we argued in the previous case, there exists an element g_j whose eigenvalue set is $\Lambda_i \cup \Lambda_i^{-1}$ and g is similar to $g_1 \oplus \cdots \oplus g_t$.

It remains to assume that $2m/k$ is even. In this case, $k = (2m, e) = (m, e) = t$. Fix $1 \leqslant j \leqslant t$ and let $\lambda_j \in \Lambda_j$. Lemma 5.3.3 implies that there exists an irreducible element $g_j \in \mathrm{SO}^-_{2m/t}(q)$ with eigenvalue set Λ_j. Lemma 2.3.7 now implies that g is similar to $g_1 \oplus \cdots \oplus g_t$, completing the proof. □

We conclude with a comment on centralisers.

Lemma 5.3.18 *Let G be $\mathrm{PGSp}_{2m}(q)$ or $\mathrm{PDO}^\varepsilon_{2m}(q)$. Let $g \in G$ lift to an element of type ${}^*(2m)^\varepsilon_q$, where $*$ is the empty symbol, Δ (q odd) or Σ (q odd and $G = \mathrm{PDO}^\varepsilon_{2m}(q)$). Then $|C_G(g)| \leqslant q^m - \varepsilon$.*

Proof A suitable power h of g has type $(2m)^\varepsilon_q$. For $x \in \mathrm{GL}_{2m}(q)$, write $\overline{x}$ for the image in $\mathrm{PGL}_{2m}(q)$. First assume that $\varepsilon = +$. Then $h = h_1 \oplus h_1^{-\mathsf{T}}$ and $|h| = \Phi^*_m(q)$. By [20, Appendix B], $|C_G(\overline{g})| \leqslant |C_G(\overline{h})| = q^m - 1$.

Next assume that $\varepsilon = -$. If $m > 1$ or q is not Mersenne, then $|h| = \Phi^*_{2m}(q)$ and from [20, Appendix B], $|C_G(\overline{g})| \leqslant |C_G(\overline{h})| = q^m + 1$. It is straightforward to verify the special case where $|h| = q + 1$ and G is $\mathrm{PGSp}_2(q)$ or $\mathrm{PDO}^-_2(q)$. □

5.3.3 *Reflections*

We conclude this section by discussing reflections. We continue to write $V = \mathbb{F}_q^{2m}$ and $\mathbb{F}_q^\times = \langle \alpha \rangle$. The standard bases $\mathcal{B}^+$ and $\mathcal{B}^-$ were introduced in (2.7) and (2.8). Recall that if q is odd, then $\beta \in \mathbb{F}_q^\times$ has order $(q-1)_2$, so $\beta \notin (\mathbb{F}_q^\times)^2$. If $\varepsilon = -$, then we will make use of the isomorphism $\Psi \colon \langle X_{r\varphi^f}, r \rangle \to \mathrm{PGO}^-_{2m}(q)$ (see Lemma 2.5.4).

Definition 5.3.19 With respect to the basis $\mathcal{B}^\varepsilon$ for $\mathbb{F}_q^2$, define

$$r^\varepsilon = \begin{pmatrix} 0 & 1 \\ 1 & 0 \end{pmatrix} \in \mathrm{O}^\varepsilon_2(q)$$

and if q is odd, then also

$$^{\Delta}r_q^+ = \begin{pmatrix} 0 & \beta \\ 1 & 0 \end{pmatrix} \in \mathrm{GO}_2^+(q)$$

and, for $\beta_2 \in \mathbb{F}_{q^2}^\times$ of order $(q^2-1)_2$,

$$^{\Delta}r_q^- = R\Psi \in \mathrm{GO}_2^-(q) \quad \text{where} \quad R = \begin{pmatrix} 0 & \beta_2 \\ \beta_2^q & 0 \end{pmatrix} \in \mathrm{GO}_2^+(q^2).$$

Lemma 5.3.20 *Let q be even and let F be an extension of $\mathbb{F}_q$. Then r^ε is a reflection that stabilises a unique (nonsingular) 1-space of F^2.*

Proof Evidently r^ε stabilises the nonsingular 1-space $\langle e_1 + f_1\rangle$ if $\varepsilon = +$ and $\langle u_1 + v_1\rangle$ if $\varepsilon = -$, and this is the unique 1-subspace stabilised by r^ε. □

Lemma 5.3.21 *Let q be odd and let F be a finite extension of $\mathbb{F}_q$. Then*

(i) *r^+ is a reflection in a vector of norm -2*
(ii) *r^- is a reflection in a vector of norm $-2\lambda^2$ for some $\lambda \in \mathbb{F}_q^\times$*
(iii) *r^ε stabilises exactly two (orthogonal nondegenerate) 1-spaces of F^2*
(iv) *$^{\Delta}r^\varepsilon$ acts irreducibly on F^2 if $|F : \mathbb{F}_q|$ is odd*
(v) *$^{\Delta}r^\varepsilon$ stabilises exactly two (orthogonal nondegenerate) 1-spaces of F^2 if $|F : \mathbb{F}_q|$ is even*
(vi) *$\tau(^{\Delta}r^\varepsilon) = \beta$ and $\det(^{\Delta}r^\varepsilon) = -\beta$.*

Proof Observe that $r^+ = r_{e_1 - f_1}$ and $(e_1 - f_1, e_1 - f_1) = -2$. Similarly, $r^- = r_{u_1 - v_1}$ and

$$(u_1 - v_1, u_1 - v_1) = 2 - 2(\xi^2 + \xi^{-2}) + 2 = -2(\xi - \xi^{-1})^2$$

(see the definition of $\mathcal{B}^-$ in (2.8)). This proves (i) and (ii).

For (iii), the characteristic polynomial of r^ε is $X^2 - 1$, so r^ε has a one-dimensional 1- and -1-eigenspace and these two 1-spaces are exactly the proper nonzero subspaces stabilised by r^ε. Similarly, (iv) and (v) hold since the characteristic polynomial of $^{\Delta}r^\varepsilon$ is $X^2 - \beta$.

Finally consider (vi). If $\varepsilon = +$, then this is a straightforward calculation. If $\varepsilon = -$, then we easily see that $\det(R) = -\beta_2^{q+1} = -\beta$ and Ψ is induced by conjugation, so $\det(^{\Delta}r^\varepsilon) = -\beta$. Similarly, $\tau(R) = \beta_2^{q+1} = \beta$, with respect to the standard plus-type form on $\mathbb{F}_{q^2}^{2m}$ and the definition of Ψ implies that $\tau(\hat{\delta}) = \beta$ with respect to the standard minus-type form on $\mathbb{F}_q^{2m}$. □

Remark 5.3.22 Let us comment on reflections.

(i) The element $r \in \mathrm{GO}_{2m}(\overline{\mathbb{F}}_p)$ from Definition 2.5.1 is simply $I_{2m-2} \perp r^+$, centralising $\langle e_1, \ldots, f_{m-1}\rangle \perp \langle e_m, f_m\rangle$. Additionally, $r\Psi = I_{2m-2} \perp r^-$, centralising $\langle e_1, \dots, f_{m-1}\rangle \perp \langle u_m, v_m\rangle$. Thus, we often identify r and r^ε as elements of $\mathrm{O}^\varepsilon_{2m}(q)$.

(ii) Assume that q is odd. By Lemma 5.3.21, the norm of r^ε is square if and only if $-2 \in (\mathbb{F}_q^\times)^2$. This latter condition holds if and only if

$$f \text{ is even or } p \equiv 1 \text{ or } 3 \pmod 8. \tag{5.11}$$

Therefore, $\ddot{r}^\varepsilon$ is $\ddot{r}_\square$ if (5.11) holds and $\ddot{r}^\varepsilon$ is $\ddot{r}_\boxtimes$ otherwise (our notation for the automorphisms induced by reflections was introduced at the beginning of Sect. 5.2.2).

(iii) If q is odd, then ${}^{\Delta}r^+ = \delta^+ r$ and ${}^{\Delta}r^- = \delta^- r$.

5.3.4 Field Extension Subgroups

We begin by stating [20, Lemma 5.3.2] for future reference.

Lemma 5.3.23 *Let k be a prime divisor of n, let $\pi : \mathrm{GL}_{n/k}(q^k).k \to \mathrm{GL}_n(q)$ be a field extension embedding and let $x \in \mathrm{GL}_{n/k}(q^k).k$ have prime order $r \neq p$.*

(i) *If $x \in \mathrm{GL}_{n/k}(q^k)$ and has eigenvalues $\lambda_1, \dots, \lambda_{n/k}$ over $\overline{\mathbb{F}}_p$, then $\pi(x)$ has eigenvalues $\Lambda_1 \cup \dots \cup \Lambda_{n/k}$ where $\Lambda_i = \{\lambda_i^{q^j} \mid 0 \leqslant j < k\}$.*

(ii) *If $x \notin \mathrm{GL}_{n/k}(q^k)$, then $r = k$ and each rth root of unity occurs as an eigenvalue of $\pi(x)$ with multiplicity n/k.*

Corollary 5.3.24 *Let G be $\mathrm{PSp}_{2m}(q)$ or $\mathrm{PSO}^\pm_{2m}(q)$ and let g lift to $g_1 \oplus \dots \oplus g_t \oplus I_\ell$ where $g_1, \dots, g_t$ have order r and distinct eigenvalues. Let $\varepsilon = \{+, -\}$. If $\varepsilon = -$, assume r is a primitive prime divisor of $q^{2d} - 1$ and each g_i acts irreducibly on a (minus-type) $2d$-space, and if $\varepsilon = -$, assume d is odd, r is a primitive prime divisor of $q^d - 1$ and each g_i centralises the decomposition of a (plus-type) $2d$-space into a dual pair of nonisomorphic totally singular d-spaces acting irreducibly on both.*

(i) *If d is odd, then g is not contained in the base of a subgroup of type $\mathrm{Sp}_m(q^2)$ (m is even) or $\mathrm{O}^\upsilon_m(q^2)$ ($\upsilon \in \{+, -\}$ if m is even and $\upsilon = \circ$ if m is odd).*

(ii) *If $\varepsilon \neq (-)^d$, then g is not contained in the base of a $\mathrm{GU}_m(q)$ subgroup.*

Proof Let $\pi : H = B.2 \to G$ be the field extension embedding in question, where B is the base of H. For a contradiction, suppose that $g \in B$.

First assume $\varepsilon = +$, so we assume d is odd. Let Λ be the nontrivial eigenvalue set of g. If $g = \pi(x)$ for $x \in B$, then, by Lemma 5.3.23(i), $\Lambda = \Lambda_0 \cup \Lambda_0^q$, where Λ_0 is the set of eigenvalues of x. Since x is defined over $\mathbb{F}_{q^2}$ we know that $\Lambda_0^{q^2} = \Lambda_0$. However, the elements of Λ_0 have order r. Since d is odd, $\Lambda_0^{q^2} = \Lambda_0^q$, so every eigenvalue of g has multiplicity at least two, contradicting the distinctness of the eigenvalues of g.

Next assume $\varepsilon = -$. Let Λ_i be the set of $2d$ distinct eigenvalues of g_i. For now consider part (i), so we may assume that d is odd. Then there are two $\mu \mapsto \mu^{q^2}$ orbits on Λ_i, say Λ_{i1} and $\Lambda_{i2} = \Lambda_{i1}^q = \Lambda_{i1}^{-1}$. By Lemma 5.3.23(i), without loss of generality, the eigenvalues of g as an element of $\mathrm{GL}_m(q^2)$ are $\bigcup_{i=1}^t \Lambda_{i1}$, which is not closed under inversion (see [20, Lemma 3.1.13]), which contradicts [20, Lemma 3.4.1].

Now consider part (ii) (still with $\varepsilon = -$). We now assume d is even. Again write Λ_{i1} and $\Lambda_{i2} = \Lambda_{i1}^{-q}$ for the two $\mu \mapsto \mu^{q^2}$ orbits on Λ_i. Then, by Lemma 5.3.23(i), without loss of generality, the eigenvalues of g as an element of $\mathrm{GU}_m(q)$ are $\bigcup_{i=1}^t \Lambda_{i1}$, which is not closed under the map $\mu \mapsto \mu^{-q}$, contradicting [20, Proposition 3.3.1]. □

Combining Corollary 5.3.24 with Lemma 5.3.17 gives the following.

Corollary 5.3.25 *Let g in $\mathrm{PSp}_{2m}(q)$ or $\mathrm{PSO}^{\pm}_{2m}(q)$ have type $(2d)^{\eta}_{q_0} \perp I_\ell$ for $q_0^e = q$.*

(i) *If d is odd, then g is not contained in the base of a subgroup of type $\mathrm{Sp}_m(q^2).2$ or $\mathrm{O}^{\upsilon}_m(q^2).2$.*
(ii) *If d is odd and $\eta = +$; or d is even, $\eta = -$ and e is odd; or d is odd, $\eta = -$ and e is even, then g is not contained in the base of a $\mathrm{GU}_m(q)$ subgroup.*

5.4 Case I: Semilinear Automorphisms

Having established the cases to consider, we now start proving Theorems 5.1 and 5.2. In this section, we begin with Case I. Accordingly, write $G = \langle T, \theta \rangle$ where $T = \mathrm{P}\Omega^{\varepsilon}_{2m}(q)$ for $m \geqslant 4$ and $\theta \in \mathrm{P\Gamma O}^{\varepsilon}_{2m}(q) \setminus \mathrm{PGO}^{\varepsilon}_{2m}(q)$. Recall the cases

(a) $G \cap \mathrm{PGO}^{\varepsilon}_{2m}(q) \leqslant \mathrm{PDO}^{\varepsilon}_{2m}(q)$
(b) $G \cap \mathrm{PGO}^{\varepsilon}_{2m}(q) \nleqslant \mathrm{PDO}^{\varepsilon}_{2m}(q)$.

The main motivation for this case distinction is that Shintani descent applies directly in Case I(a) but in Case I(b) we need to use this technique in a more flexible manner. A side effect of this distinction is that in Case I(a), $\nu(x) > 1$ for all $x \in G \cap \mathrm{PGL}(V)$ and this makes the probabilistic method easier to apply. Recall that Fig. 5.1 further partitions Cases I(a) and I(b). In particular, I(a) is the union of I(i), (ii) and (iv), and I(b) is the union of I(iii) and (v), where the definitions of Cases I(i)–(v) are summarised in Fig. 5.2. We consider Cases I(a) and I(b) in Sects. 5.4.1 and 5.4.2, respectively.

5.4.1 *Case I(a)*

In this section, we first identify an element $t\theta \in G$, then we determine $\mathcal{M}(G, t\theta)$ and apply the probabilistic method discussed in Sect. 2.1.

Shintani descent (see Chap. 3) will play an indispensable role in identifying an appropriate element $t\theta \in T\theta$ for each automorphism θ (see Example 3.1.5). With this in mind let us fix the following notation for Sect. 5.4.1.

Notation 5.4.1 Write $q = p^f$ where $f \geqslant 2$. Let $V = \mathbb{F}_q^{2m}$ with $m \geqslant 4$. Fix the simple algebraic group

$$X = \mathrm{PDO}_{2m}(\overline{\mathbb{F}}_p) = \begin{cases} \Omega_{2m}(\overline{\mathbb{F}}_2) & \text{if } p = 2 \\ \mathrm{PSO}_{2m}(\overline{\mathbb{F}}_p) & \text{if } p \text{ is odd.} \end{cases}$$

Fix the standard Frobenius endomorphism $\varphi = \varphi_{\mathcal{B}^+}$ of X, defined with respect to the standard basis $\mathcal{B}^+$, as $(a_{ij}) \mapsto (a_{ij}^p)$, modulo scalars.
Fix the diagonal element δ^+ and reflection r (see Definitions 5.2.2 and 2.5.1).
If $\varepsilon = -$, fix the map Ψ from Lemma 2.5.4, which restricts to an isomorphism $\Psi\colon \langle X_{r\varphi^f}, r\rangle \to \mathrm{PGO}^-_{2m}(q)$. Moreover, fix $\psi = \Psi \circ \varphi \circ \Psi^{-1}$ and $\delta^- = \delta^+\Psi$ (see (5.7) and Definition 5.2.8).

As a consequence of Proposition 5.2.15, we can assume that $\theta \in \mathrm{PGO}^+_{2m}(q)\varphi^i$ when $\varepsilon = +$ and $\theta \in \mathrm{PGO}^-_{2m}(q)\psi^i$ when $\varepsilon = -$. In the latter case, the definition of Case I(a) ensures that $2f/i$ is odd, so i is even and it is straightforward to show, for $j = i/2$, we have $2f/(2f, f+j) = 2f/(2f, i)$. Consequently, when $\varepsilon = -$, we may, and will, work with

$$\theta = \theta_0\psi^{f+j} = \theta_0 r\psi^j$$

instead of $\theta_0\psi^i$, noting that j divides f and $2f/i = f/j$ is odd.

Fig. 5.2 Definition of Cases I(i)–(v)

case	ε	θ	condition
(i)	+	$\theta_0\varphi^i$	none
(ii)		$\theta_0 r\varphi^i$	f/i is even
(iii)			f/i is odd
(iv)	–	$\theta_0\psi^i$	$2f/i$ is odd
(v)			$2f/i$ is even

Note: $\theta_0 \in \mathrm{Inndiag}(T)$

Notation 5.4.1. (continued)
Write $q = q_0^e$, where (η, σ, e) are as follows

case	η	σ	e
(i)	$+$	φ^i	f/i
(ii)	$-$	$r\varphi^i$	f/i
(iv)	$-$	$r\varphi^j$	$2f/i$

Let F be the Shintani map of (X, σ, e), so

$$F: \{(g\widetilde{\sigma})^{X_{\sigma^e}} \mid g \in X_{\sigma^e}\} \to \{x^{X_\sigma} \mid x \in X_\sigma\}.$$

Observe that $X_{\sigma^e} \cong \mathrm{Inndiag}(T)$ and $X_\sigma = \mathrm{Inndiag}(T_0)$ for a subgroup T_0 of T isomorphic to $\mathrm{P}\Omega^\eta_{2m}(q_0)$. We will harmlessly identify T_0 with $\mathrm{P}\Omega^\eta_{2m}(q_0)$ and write $\mathrm{Inndiag}(T_0) = \mathrm{PDO}^\eta_{2m}(q_0) = \langle \mathrm{PSO}^\eta_{2m}(q_0), \delta_0\rangle$.

Remark 5.4.2 Let us make some observations regarding Notation 5.4.1.

(i) The definition of Case I(a) implies that $\varepsilon = \eta^e$.
(ii) If $\varepsilon = +$, then $\mathrm{Inndiag}(T)\theta = X_{\sigma^e}\widetilde{\sigma}$.
(iii) If $\varepsilon = -$, then, via the isomorphism Ψ, we identify X_{σ^e} with $\mathrm{Inndiag}(T)$ and we identify $\widetilde{\sigma} = r\varphi^j$ with $\theta = r\psi^j$, so $\mathrm{Inndiag}(T)\theta = X_{\sigma^e}\widetilde{\sigma}$ in this case also.

In light of Remark 5.4.2, the main idea is to select the element $t\theta \in \mathrm{Inndiag}(T)\sigma$ as the preimage under F of a carefully chosen element $y \in \mathrm{Inndiag}(T_0)$. If q is even, then $\mathrm{Inndiag}(T) = T$ and this is a transparent process. When q is odd, the following two results facilitate this selection procedure (compare with Example 3.2.4.)

Lemma 5.4.3 *Let q be odd. The Shintani map F restricts to bijections*

(i) $F_1\colon \{(g\widetilde{\sigma})^{\mathrm{PDO}^\varepsilon_{2m}(q)} \mid g \in \mathrm{PSO}^\varepsilon_{2m}(q)\} \to \{x^{\mathrm{PDO}^\eta_{2m}(q_0)} \mid x \in \mathrm{PSO}^\eta_{2m}(q_0)\}$
(ii) $F_2\colon \{(g\delta\widetilde{\sigma})^{\mathrm{PDO}^\varepsilon_{2m}(q)} \mid g \in \mathrm{PSO}^\varepsilon_{2m}(q)\} \to \{(x\delta_0)^{\mathrm{PDO}^\eta_{2m}(q_0)} \mid x \in \mathrm{PSO}^\eta_{2m}(q_0)\}$.

Proof This is Lemma 3.2.2 with $\pi\colon \mathrm{SO}_{2m}(\overline{\mathbb{F}}_q) \to \mathrm{PSO}_{2m}(\overline{\mathbb{F}}_q)$, noting that $\langle \mathrm{PSO}^\varepsilon_{2m}(q), \widetilde{\sigma}\rangle$ and $\mathrm{PSO}^\eta_{2m}(q_0)$ are index two subgroups of $\langle \mathrm{PDO}^\varepsilon_{2m}(q), \widetilde{\sigma}\rangle$ and $\mathrm{PDO}^\eta_{2m}(q_0)$. □

Lemma 5.4.4 *Let q be odd and assume that $q_0^m \equiv \eta \pmod 4$. The map F_1 restricts to bijections*

(i) $F_{11}\colon \{(g\widetilde{\sigma})^{\mathrm{PDO}^\varepsilon_{2m}(q)} \mid g \in T\} \to \{x^{\mathrm{PDO}^\eta_{2m}(q_0)} \mid x \in T_0\}$
(ii) $F_{12}\colon \{(gr_\square r_\boxtimes\widetilde{\sigma})^{\mathrm{PDO}^\varepsilon_{2m}(q)} \mid g \in T\} \to \{(xr_\square r_\boxtimes)^{\mathrm{PDO}^\eta_{2m}(q_0)} \mid x \in T_0\}$.

Proof The condition $q_0^m \equiv \eta \pmod 4$ ensures that $|\,\mathrm{PSO}^{\eta}_{2m}(q_0) : T_0| = 2$ (see (2.2)). We claim $|\,\mathrm{PSO}^{\varepsilon}_{2m}(q) : T| = 2$. If $\varepsilon = \eta = +$, then $q^m \equiv 1 \pmod 4$ and $|\,\mathrm{PSO}^{+}_{2m}(q) : T| = 2$. Next, if $\varepsilon = +$ and $\eta = -$, then e is even, so again $q^m \equiv 1 \pmod 4$ and $|\,\mathrm{PSO}^{+}_{2m}(q) : T| = 2$. Finally, if $\varepsilon = \eta = -$, then e is odd and $q^m \equiv 3 \pmod 4$, so $|\,\mathrm{PSO}^{-}_{2m}(q) : T| = 2$.

Write $W = \mathrm{Spin}_{2m}(\overline{\mathbb{F}}_p)$ and let $\pi : W \to X$ be the natural isogeny. Now $\pi(W_{\sigma^e}) = T$ where $W_{\sigma^e} = \mathrm{Spin}^{\varepsilon}_{2m}(q)$, and $\pi(W_\sigma) = T_0$ where $W_\sigma = \mathrm{Spin}^{\eta}_{2m}(q_0)$ (see [37, Theorem 2.2.6(f)]). Evidently, $T_0 \trianglelefteq \mathrm{Inndiag}(T_0)$. Moreover, if $\varepsilon = +$, then the condition $q_0^m \equiv \eta \pmod 4$ implies that condition (5.6) is satisfied, so, in light of Remark 5.2.4, $\langle \ddot{\sigma} \rangle \trianglelefteq \langle \mathrm{Inndiag}(T)/T, \ddot{\sigma} \rangle$ and hence $\langle T, \widetilde{\sigma} \rangle \trianglelefteq \langle \mathrm{Inndiag}(T), \widetilde{\sigma} \rangle$. Similarly, if $\varepsilon = -$, then i is even, so $[\ddot{\psi}^i, \ddot{\delta}] = 1$ (see Lemma 5.2.12), which implies that $\langle \ddot{\sigma} \rangle \trianglelefteq \langle \mathrm{Inndiag}(T)/T, \ddot{\sigma} \rangle$ and hence, again, $\langle T, \widetilde{\sigma} \rangle \trianglelefteq \langle \mathrm{Inndiag}(T), \widetilde{\sigma} \rangle$. Therefore, by Lemma 3.2.2, the Shintani map F of (X, σ, e) restricts to the map F_{11}. By Lemma 5.4.3, F restricts to F_1, so, in fact, F_1 restricts to the bijections F_{11} and F_{12}, as required. □

We will now define the elements we will use to prove Theorems 5.1 and 5.2 in Case I(a). In light of the probabilistic method outlined in Sect. 2.1, we need to select $t\theta \in G$ in a way which allows us to control both the maximal subgroups of G which contain it and the fixed point ratios associated with these subgroups.

Recall that in Definitions 5.3.7 and 5.3.9, we defined standard *types* of elements denoted $(2d)^{\pm}_q$ for some $d \geqslant 1$. Moreover, in Definitions 5.3.13 and 5.3.15, for odd q we also defined variants indicated by superscripts Δ and Σ. These variants have a very similar action on the natural module but crucially are contained in a different coset of the simple group. By working with the latter, we will be able to select an element that lies in the precise coset $T\theta$.

In the following definition, the final case in each definition is the empty symbol.

Definition 5.4.5 Let $\theta \in \mathrm{Aut}(T)$.

(i) Define

$$a = a(\theta) = \begin{cases} \Delta \text{ if } \theta \notin \langle \mathrm{PO}^{\varepsilon}_{2m}(q), \varphi \rangle \\ \quad\text{if } \theta \in \langle \mathrm{PO}^{\varepsilon}_{2m}(q), \varphi \rangle \end{cases}$$

(ii) Define

$$b = b(\theta) = \begin{cases} \Delta \text{ if } \theta \notin \langle \mathrm{PO}^{\varepsilon}_{2m}(q), \varphi \rangle \\ \Sigma \text{ if } \theta \in \langle \mathrm{P\Omega}^{\varepsilon}_{2m}(q), \varphi \rangle r_{\square} r_{\boxtimes} \\ \quad\text{otherwise.} \end{cases}$$

(iii) Define

$$c = c(\theta, q_0) = \begin{cases} \Delta \text{ if } \theta \notin \langle \mathrm{PO}^{\varepsilon}_{2m}(q), \varphi \rangle \\ \Sigma \text{ if } \theta \in \langle \mathrm{P\Omega}^{\varepsilon}_{2m}(q), \varphi \rangle \text{ and } q_0 \text{ is Mersenne} \\ \Sigma \text{ if } \theta \in \langle \mathrm{P\Omega}^{\varepsilon}_{2m}(q), \varphi \rangle r_{\square} r_{\boxtimes} \text{ and } q_0 \text{ is not Mersenne} \\ \quad\text{otherwise.} \end{cases}$$

Generic case		
m	y	
(mod 4)	$\eta = +$	$\eta = -$
0	${}^{c}(m)^{-} \perp {}^{a}(m-2)^{+} \perp {}^{a}(2)^{-}$	${}^{c}(2m-2)^{+} \perp {}^{a}(2)^{-}$
2	${}^{c}(m)^{+} \perp {}^{a}(m-2)^{-} \perp {}^{a}(2)^{-}$	
3	${}^{c}(2m-2)^{-} \perp {}^{a}(2)^{-}$	${}^{c}(m+1)^{-} \perp {}^{a}(m-3)^{-} \perp {}^{a}(2)^{-}$
1		${}^{c}(m+3)^{-} \perp {}^{a}(m-5)^{-} \perp {}^{a}(2)^{-}$

Specific cases		
m	y	
	$\eta = +$	$\eta = -$
4	${}^{c}(6)^{-} \perp {}^{a}(2)^{-}$	${}^{b}(8)^{-}$
5		${}^{b}(6)^{+} \perp {}^{a}(4)^{-}$

Note: we describe y by specifying its type over $\mathbb{F}_{q_0}$

Fig. 5.3 Case I(a): The element y for the automorphism θ

Remark 5.4.6 The dependence on whether q_0 is Mersenne in Lemma 5.3.11 has to be taken into account in our arguments and defining c as a variant on b that depends on q_0 allows us to do this. Notice that $a = b = c$ is empty when q is even.

Proposition 5.4.7 *Let $T = \mathrm{P}\Omega^{\varepsilon}_{2m}(q)$ and let θ be an automorphism in Fig. 5.1 (in Case I(i), I(ii) or I(iv)). Let $y \in \mathrm{PDO}^{\eta}_{2m}(q_0)$ be the element in Fig. 5.3. Then there exists $t \in T$ such that $(t\theta)^e$ is X-conjugate to y.*

Proof As $y \in \mathrm{PDO}^{\eta}_{2m}(q_0)$, by Theorem 3.1.4, there exists $g \in \mathrm{Inndiag}(T)$ such that $(g\widetilde{\sigma})^e$ is X-conjugate to y. We will now prove that $g\widetilde{\sigma}$ is contained in the coset $T\theta$. It is routine to deduce information about which coset of T_0 contains y. For example, assume that m is even and $\eta = -$. If q is even, then y has type $(2m-2)^{+}_{q_0} \perp (2)^{-}_{q_0}$ and y is clearly an element of T_0. Now assume that q is odd and fix $\widehat{y} = \widehat{y}_1 \perp \widehat{y}_2 \in \mathrm{DO}^{-}_{2m}(q_0)$ where $\widehat{y}_1$ has type ${}^{b}(2m-2)^{+}_{q_0}$ and $\widehat{y}_2$ has type ${}^{a}(2)^{-}_{q_0}$ such that $y = \widehat{y}Z(\mathrm{DO}^{-}_{2m}(q_0))$. If $\theta \in \{\delta, r_{\square}r_{\boxtimes}\delta\}$, then $\widehat{y}$ has type ${}^{\Delta}(2m-2)^{+}_{q_0} \perp {}^{\Delta}(2)^{-}_{q_0}$, so $\tau(\widehat{y}_1) = \tau(\widehat{y}_2) = \beta$ and we deduce that $y \in \mathrm{PSO}^{+}_{2m}(q_0)\delta_0$. For now assume that q_0 is not Mersenne. If $\theta = 1$, then $\widehat{y}$ has type $(2m-2)^{+}_{q_0} \perp (2)^{-}_{q_0}$, so, by Lemma 5.3.11, $y \in T_0$, and if $\theta = r_{\square}r_{\boxtimes}$, then $\widehat{y}$ has type ${}^{\Sigma}(2m-2)^{+}_{q_0} \perp (2)^{-}_{q_0}$, so Lemma 5.3.11 implies that $(2)^{-}_{q_0} \in \Omega^{-}_{2}(q_0)$ and Lemma 5.3.14 implies that ${}^{\Sigma}(2m-2)^{+}_{q_0} \in \mathrm{SO}^{+}_{2m-2}(q_0)$, so $y \in T_0 r_{\square}r_{\boxtimes}$. Now assume that q_0 is Mersenne. If $\theta = 1$, then $\widehat{y}$ has type ${}^{\Sigma}(2m-2)^{+}_{q_0} \perp (2)^{-}_{q_0}$, so $\widehat{y}_1 \in \mathrm{SO}^{+}_{2m-2}(q_0) \setminus \Omega^{+}_{2m-2}(q_0)$ and $\widehat{y}_2 \in \mathrm{SO}^{-}_{2m}(q_0) \setminus \Omega^{-}_{2m}(q_0)$ and therefore $y \in \Omega^{-}_{2m}(q_0)$. Similarly, if $\theta = r_{\square}r_{\boxtimes}$, then $\widehat{y}$ has type $(2m-2)^{+}_{q_0} \perp (2)^{-}_{q_0}$ and we deduce that $y \in \mathrm{SO}^{-}_{2m}(q_0) \setminus \Omega^{-}_{2m}(q_0)$.

We will now use Shintani descent (in particular Lemmas 5.4.3 and 5.4.4) to determine the coset of T containing $g\widetilde{\sigma}$.

If q is even, then $\widetilde{\sigma} = \theta$ (one of φ^i, $r\varphi^i$ and ψ^i) and $X_{\sigma^e} = T$, so $g\widetilde{\sigma} \in T\theta$.

Therefore, from now on we may assume that q is odd. Assume that θ appears in Row (2) of Fig. 5.1. Then $\tau(y) = \beta_0$, so $y \in \mathrm{PSO}^{\eta}_{2m}(q_0)\delta_0$. By Lemma 5.4.3, this implies that $g\widetilde{\sigma} \in \mathrm{PSO}^{\varepsilon}_{2m}(q)\delta\widetilde{\sigma}$. Therefore, $g\widetilde{\sigma} = t\theta$ where $t \in T$ and $\theta \in \{\delta\widetilde{\sigma}, r_{\square}r_{\boxtimes}\delta\widetilde{\sigma}\}$. In Case I(i), $\theta = \delta\varphi^i$, in Case I(ii) $\theta = \delta r\varphi^i$ and in Case I(iv) $\theta = \delta\psi^i$, which suffices to prove the claim, since in all three cases, $\ddot{\theta}$ and $\ddot{r}_{\square}\ddot{r}_{\boxtimes}\ddot{\theta}$ are Out(T)-conjugate (see Lemmas 5.2.6 and 5.2.14).

Now assume θ appears in Row (1) or (3). Then $\tau(y) = 1$, so $y \in \mathrm{PSO}^{\eta}_{2m}(q_0)$ and $g\widetilde{\sigma} \in \mathrm{PSO}^{\varepsilon}_{2m}(q)\widetilde{\sigma}$, by Lemma 5.4.3. If $D(Q) = \boxtimes$, then $\widetilde{\sigma} = \theta$ (one of φ^i, $r\varphi^i$ and ψ^i) and $T = \mathrm{PSO}^{\varepsilon}_{2m}(q)$, so $g\widetilde{\sigma} \in T\theta$.

Therefore, it remains to assume that $D(Q) = \square$. In this case, $q^m \equiv \varepsilon \pmod 4$. For now assume that $q_0^m \equiv \eta \pmod 4$, so that we may apply Lemma 5.4.4 (this always holds when $\varepsilon = -$). By the choice of a and c, if θ is in Row (1), then the spinor norm of y is square, so $y \in \mathrm{P\Omega}^{\eta}_{2m}(q_0)$ and $g\widetilde{\sigma} \in T\widetilde{\sigma}$, by Lemma 5.4.4, and, since $\theta = \widetilde{\sigma}$ (one of φ^i, $r\varphi^i$ and ψ^i), we conclude that $g\widetilde{\sigma} \in T\theta$. Similarly, if θ is in Row (3), then $y \in \mathrm{PSO}^{\eta}_{2m}(q_0) \setminus \mathrm{P\Omega}^{\eta}_{2m}(q_0)$ and $g\widetilde{\sigma} \in Tr_{\square}r_{\boxtimes}\widetilde{\sigma}$, so $g\widetilde{\sigma} \in T\theta$ since $\theta = \widetilde{\sigma}$ (one of $r_{\square}r_{\boxtimes}\varphi^i$, $r_{\square}r_{\boxtimes}r\varphi^i$ or $r_{\square}r_{\boxtimes}\psi^i$).

We now need to assume that $q^m \equiv \varepsilon \pmod 4$ but $q_0^m \not\equiv \eta \pmod 4$. In this case $\varepsilon = +$. First assume that $\eta = +$. Therefore, $q_0 \equiv 3 \pmod 4$ and m is odd. This forces $q \equiv 1 \pmod 4$. Together this implies that m is odd, $p \equiv 3 \pmod 4$, i is odd, f is even. Under these conditions, we need only consider one of φ^i and $r_{\square}r_{\boxtimes}\varphi^i$ (see Remarks 5.2.16 and 5.2.16), so we can choose θ such that $g\widetilde{\sigma} \in T\theta$. Now assume that $\eta = -$. Therefore, $q_0 \equiv 1 \pmod 4$, so m is even or i is even or $p \equiv 1 \pmod 4$. This allows us to only consider one of $r\varphi^i$ and $r_{\square}r_{\boxtimes}r\varphi^i$ (see Remarks 5.2.16 and 5.2.16), so, as above, we can choose θ such that $g\widetilde{\sigma} \in T\theta$. This completes the proof. $\square$

Continue to let T be the simple group $\mathrm{P\Omega}^{\varepsilon}_{2m}(q)$ and let θ be an automorphism from Fig. 5.1. Fix $y \in \mathrm{PDO}^{\eta}_{2m}(q_0)$ from Fig. 5.3 and $t\theta \in G = \langle T, \theta \rangle$ from Proposition 5.4.7. We will now study the set $\mathcal{M}(G, t\theta)$ of maximal overgroups of $t\theta$ in G. For now we will assume that $T \neq \mathrm{P\Omega}^{\varepsilon}_{8}(q)$ and we will return to this special case at the end of the section.

Theorem 5.4.8 *Assume that $T \neq \mathrm{P\Omega}^{\varepsilon}_{8}(q)$. The maximal subgroups of G which contain $t\theta$ are listed in Fig. 5.4, where $m(H)$ is an upper bound on the number of subgroups of type H in $\mathcal{M}(G, t\theta)$.*

Let us outline the proof of Theorem 5.4.8. If $T \leqslant H$, then we deduce that $\theta \in H$, since $t\theta \in H$, but then we would have $H = G$, which is not the case. Therefore, $T \not\leqslant H$, so Theorem 2.6.1 implies that H is contained in one of the geometric families $\mathcal{C}_1, \dots, \mathcal{C}_8$ or is an almost simple irreducible group in $\mathcal{S}$.

Our approach is to consider each of these families in turn and determine which possible types of subgroup could contain the element $t\theta$, by exploiting the restrictive properties that we have chosen $t\theta$ to have. For types of subgroups which could contain the element $t\theta$ we will find an upper bound on the number of subgroups of this type that contain $t\theta$. We will not concern ourselves with determining *exactly*

	type of H	$m(H)$	conditions
$\mathcal{C}_1$	$\mathrm{O}_2^{\upsilon}(q) \times \mathrm{O}_{2m-2}^{\varepsilon\upsilon}(q)$	1	$(\eta, m) \neq (-, 5)$
	$P_{m/2}$	2	$\eta = +, \ m \equiv 0 \pmod 4$
	$P_{m/2-1}$	2	$\eta = +, \ m \equiv 2 \pmod 4$
	$\mathrm{O}_{m-2}^{\upsilon}(q) \times \mathrm{O}_{m+2}^{\varepsilon\upsilon}(q)$	1	$\eta = +, \ m$ even
	P_{m-1}	2	$\eta = -, \ m$ even
	$\mathrm{O}_{m-3}^{\upsilon}(q) \times \mathrm{O}_{m+3}^{\varepsilon\upsilon}(q)$	1	$\eta = -, \ m$ odd
	$\mathrm{O}_{m-5}^{\upsilon}(q) \times \mathrm{O}_{m+5}^{\varepsilon\upsilon}(q)$	1	$\eta = -, \ m \equiv 1 \pmod 4), \ m \neq 5$
	$\mathrm{O}_{m-1}^{\upsilon}(q) \times \mathrm{O}_{m+1}^{\varepsilon\upsilon}(q)$	1	$\eta = -, \ m \equiv 3 \pmod 4$
	P_3	2	$\eta = -, \ m = 5$
	$\mathrm{O}_4^{\upsilon}(q) \times \mathrm{O}_6^{\varepsilon\upsilon}(q)$	1	$\eta = -, \ m = 5$
$\mathcal{C}_2$	$\mathrm{O}_{2m/k}^{\upsilon}(q) \wr S_k$	N	$k \mid m, \ k > 1, \ \upsilon^k = \varepsilon$
	$\mathrm{O}_{2m/k}(q) \wr S_k$	N	$k \mid 2m, \ 2m/k > 1$ odd
	$\mathrm{GL}_m(q)$	$2N$	$\eta = +, \ m$ even
		N	$\varepsilon = +, \ \eta = -, \ m$ odd
$\mathcal{C}_3$	$\mathrm{O}_m(q^2)$	$2N$	$m > 5$ odd
	$\mathrm{GU}_m(q)$	$2N$	$\varepsilon = \eta = +, \ m$ even
		N	$\varepsilon = \eta = -, \ m$ odd
$\mathcal{C}_5$	$\mathrm{O}_{2m}^{\upsilon}(q^{1/k})$	N	$k \mid f, \ k$ is prime, $\upsilon^k = \varepsilon$
$\mathcal{S}$	$\mathrm{PSp}_4(q)$	$2N$	$\eta = -, \ m = 5, \ q \equiv \varepsilon \pmod 4$

Note: $N = |C_{\mathrm{PDO}_{2m}^{\eta}(q_0)}(y)|$ and in $\mathcal{C}_1$ there is a unique choice of υ

Fig. 5.4 Case I(a): Description of $\mathcal{M}(G, t\theta)$

which subgroups contain $t\theta$; sometimes it will be sufficient, for example, to use an overestimate on the number of subgroups of a given type which contain $t\theta$.

Remark 5.4.9 If $s \in \mathrm{GL}_n(q)$ has order divisible by a primitive prime divisor of $q^k - 1$ for $k > \frac{n}{2}$, then the subgroups $H \leqslant \mathrm{GL}_n(q)$ that contain s are classified by the main theorem of [40]. However, this will not be useful in proving Theorem 5.4.8. To see why, consider the example where $G = \langle \mathrm{P}\Omega_{2m}^{+}(p^f), \varphi \rangle$ and φ is the standard order f field automorphism. Then, via Shintani descent, we choose an element $t\varphi \in G$ such that $(t\varphi)^f$ is conjugate to an element of $\mathrm{P}\Omega_{2m}^{+}(p)$. Therefore, $t\varphi$, and even more so $(t\varphi)^f$, which is the element we typically have better information about, has a small order compared with the order of G. Consequently, we will need to use other properties of the element $t\varphi$ in order to constrain its maximal overgroups.

We will present a result on multiplicities of subgroups in $\mathcal{M}(G, t\theta)$, before proving Theorem 5.4.8 in three parts, by considering the cases where $H \in \mathcal{M}(G, t\theta)$ is reducible, irreducible imprimitive and primitive with respect to the action of $H \cap T$ on the natural module for T. We write

$$\widetilde{G} = \langle X_{\sigma^e}, \widetilde{\sigma} \rangle$$

noting that $\mathrm{Inndiag}(T) \leqslant \widetilde{G} \leqslant \mathrm{Aut}(T)$ and $G \leqslant \widetilde{G}$.

The following result will apply to Case I(b) also.

Proposition 5.4.10 *Assume that $T \neq \mathrm{P\Omega}_8^+(q)$. Let $T \leqslant A \leqslant \mathrm{Aut}(T)$ and let H be a maximal $\mathcal{C}_1$, $\mathcal{C}_2$, $\mathcal{C}_3$ or $\mathcal{C}_5$ subgroup of A. Then there is a unique $\widetilde{G}$-conjugacy class of subgroups of type H, unless H has one of the following types, in which case there are two $\widetilde{G}$-classes:*

type	P_m	$\mathrm{GL}_m(q)$	$\mathrm{GU}_m(q)$	$\mathrm{O}_m^+(q^2)$	$\mathrm{O}_m(q^2)$
ε	$+$	$+$	$+$	$+$	$-$
m	*any*	*odd*	*even*	*even*	*odd*

Proof If $m \leqslant 6$, then the result follows from the tables in [8, Chapter 8]. Now assume that $m \geqslant 7$. We will apply the main theorem of [50].

Let H be a maximal geometric subgroup of G. Let $\mathcal{H} = \{H_1, \dots, H_c\}$ be a set of representatives of the c distinct T-classes of subgroups of T of the same type as H. In the terminology of [50, Chapter 3], for each $1 \leqslant i \leqslant c$, let $H_{G,i}$ be the G-associate of H_i. In particular, $H_{G,i}$ is a geometric subgroup of G of the same type as H_i and $H_i \leqslant H_{G,i}$ (see [50, Section 3.1] for a precise definition). There is a natural action of $\mathrm{Out}(T)$ on the set $\mathcal{H}$, and the permutation representation $\pi : \mathrm{Out}(T) \to S_c$ associated to this action is described in [50, Tables 3.5.A–3.5.G]. As a consequence of the proof of [50, Lemma 3.2.2(iii)], for $G \leqslant A \leqslant \mathrm{Aut}(T)$, the groups $H_{G,i}$ and $H_{G,j}$ are A-conjugate if and only if H_i and H_j are in the same $\pi(\ddot{A})$-orbit. By [50, Tables 3.5E and 3.5G], $\pi(\widetilde{G}/T)$ is transitive, except for the exceptional cases in the statement, when $c = 2$ and $\pi(\widetilde{G}/T)$ is intransitive. This proves the statement, but we provide some examples, with $\varepsilon = +$.

For example, consider the case where m is odd, H has type $\mathrm{O}_{2m}^-(q^{1/2})$ and $p \equiv 1 \pmod 4$. In this situation, $c = 4$, $\ker(\pi) = \langle \ddot{\varphi} \rangle$ and the stabiliser of H_1 is $\langle \ddot{\varphi}, \ddot{r}_\square \rangle$. Therefore, $\pi(\widetilde{G}/T) = \langle \ddot{\delta} \rangle \cong C_4$ is transitive, so there is exactly one $\widetilde{G}$-class of subgroups of G of the same type as H.

For another example, let m be even and let H have type $\mathrm{GL}_m(q)$. In this situation, $c = 2$, $\ker(\pi) = \langle \mathrm{Inndiag}(T)/T, \ddot{\varphi} \rangle$ and the stabiliser of H_1 is $\langle \ddot{\varphi}, \ddot{r}_\square \rangle$. Therefore, $\pi(\widetilde{G}/T) = 1$, so there are exactly two $\widetilde{G}$-classes of subgroups of G of the same type as H. □

Proposition 5.4.11 *Theorem 5.4.8 is true for reducible subgroups.*

Proof We will apply Lemma 3.3.2 (see Example 3.3.3).

Case 1: Stabilisers of Totally Singular Subspaces Let H be a maximal parabolic subgroup of G. Then $H \leqslant \widetilde{H} = \langle Y_{\sigma^e}, \widetilde{\sigma} \rangle$ for a $\widetilde{\sigma}$-stable parabolic subgroup $Y \leqslant X$. In particular, Y is a closed connected subgroup of X. Moreover, $\widetilde{H}$ and Y_σ are maximal (and hence self-normalising) subgroups of $\widetilde{G}$ and X_σ, respectively.

Therefore, Lemma 3.3.2 implies that the number of X_{σ^e}-conjugates of H which contain $t\theta$ equals the number of X_σ-conjugates of $H \cap X_\sigma$ which contain $F(t\theta) = y$.

Assume that $\eta = +$ and $m \geqslant 5$ with $m \pmod 4 \in \{0, 1\}$; the other cases are similar. First assume that $m \equiv 1 \pmod 4$. By Lemma 2.3.3, y does not stabilise any totally singular subspaces of $\mathbb{F}_{q_0}^n$ and therefore is not contained in any parabolic subgroups of X_σ. Therefore, $t\theta$ is not contained in any parabolic subgroups of G.

Now assume that $m \equiv 0 \pmod 4$. Here y stabilises exactly two totally singular subspaces of $\mathbb{F}_{q_0}^n$, each of dimension $m/2 - 1$, so y is contained in exactly two parabolic subgroups of X_σ, of type $P_{m/2-1}$, and consequently $t\theta$ is contained in exactly two parabolic subgroups of G, of type $P_{m/2-1}$, as claimed in Theorem 5.4.8.

Case 2: Stabilisers of Nondegenerate Subspaces Let H be the stabiliser in G of a nondegenerate k-space. Let $L = \mathrm{SL}_n(\overline{\mathbb{F}}_p)/\langle -I_n\rangle$ and extend the domain of σ to L. Let E be the Shintani map of (L, σ, e). Observe that $t\theta \in G \leqslant \langle L_{\sigma^e}, \theta\rangle$ and $F(t\theta) \in X_\sigma \leqslant L_\sigma$. Accordingly, Lemma 3.2.1 implies that $F(t\theta) = E(t\theta)$. Let $M \leqslant L$ be a P_k parabolic subgroup. Applying Lemma 3.3.2 to the Shintani map E for L and the subgroup $M \leqslant L$, we see that the number of k-spaces of $V = \mathbb{F}_q^n$ fixed by $t\theta$ equals the number of k-spaces of $V_0 = \mathbb{F}_{q_0}^n$ fixed by $E(t\theta) = F(t\theta) = y$.

Again let us assume that $m \pmod 4 \in \{0, 1\}$ and $\eta = +$, beginning with the case where $m \equiv 1 \pmod 4$. Lemma 2.3.3 implies that y stabilises exactly two proper nonzero subspaces of V_0, of dimensions 2 and $2m - 2$, so $t\theta$ stabilises exactly two proper nonzero subspaces of V, of the same dimensions. In Case 1, we demonstrated that $t\theta$ is not contained in a parabolic subgroup of G. Therefore, both subspaces of V must be nondegenerate, for otherwise $t\theta$ would stabilise its (totally singular) radical and therefore be contained in a parabolic subgroup. Consequently, the only reducible maximal subgroup of G containing $t\theta$ has type $\mathrm{O}_2^{\upsilon}(q) \times \mathrm{O}_{2m-2}^{-\upsilon}(q)$ for some sign $\upsilon \in \{+, -\}$ (it is exactly for the reason that we pass to the linear group L that we cannot determine the sign υ).

Now assume that $m \equiv 0 \pmod 4$. Then y stabilises exactly 14 proper nonzero subspaces of V_0, of dimensions

$$2,\quad m/2-1\ (2),\quad m/2+1\ (2),\quad m-2\ (2),\quad m\ (2),$$
$$m+2,\quad 3m/2-1\ (2),\quad 3m/2+1\ (2),\quad 2m-2,$$

where (2) denotes the fact that there are two subspaces of each of these dimensions.

From Case 1, we know that $t\theta$ stabilises exactly two totally singular subspaces, each of dimension $m/2 - 1$. Since $t\theta$ stabilises a (necessarily not totally singular) 2-, m- and $(m + 2)$-space, we deduce that the stabilised $(m/2 + 1)$-, $(3m/2 - 1)$- and $(3m/2 + 1)$-spaces must be the direct sum of 2 the 2-, m- and $(m + 2)$-spaces with the two $(m/2 - 1)$-spaces. These subspaces are not totally singular, since there are only two such subspaces stabilised by $t\theta$. We now claim that these six subspace are degenerate. Indeed, the $(m/2 + 1)$-space has a $(m/2 - 1)$-dimensional totally singular subspace, which implies that it is degenerate. In addition, if one of the $(3m/2 \pm 1)$-spaces were nondegenerate, then $t\theta$ would stabilise its $(m/2 \mp 1)$-dimensional nondegenerate orthogonal complement, but we have already shown that all subspaces of this dimension stabilised by $t\theta$ are degenerate. Therefore, the only possible nondegenerate subspaces of V stabilised by $t\theta$ are those of dimension 2, $m-2$, m (of which there are two), $m+2$, $2m-2$, which give rise to maximal reducible subgroups in the statement. □

Proposition 5.4.12 *Theorem 5.4.8 is true for imprimitive subgroups.*

Proof By [50, Table 3.5.E], all possible types of irreducible imprimitive subgroup feature in Fig. 5.4. If $\varepsilon = +$, then we claim that maximal subgroups of type $\mathrm{GL}_m(q)$ only arise if $\eta = +$ and m is even, or $\eta = -$ and m is odd.

First consider $\eta = +$ and m odd. In this case, $G \leqslant \langle \mathrm{Inndiag}(T), \varphi^i \rangle$, so there are no elements in G which interchange the totally singular subspaces $\langle e_1, \ldots, e_m \rangle$ and $\langle f_1, \ldots, f_m \rangle$ (see [50, Proposition 2.7.4]). Therefore, a subgroup of G of type $\mathrm{GL}_m(q)$ is contained in two subgroups of type of P_m, and no maximal subgroups of type $\mathrm{GL}_m(q)$ occur.

Now consider $\eta = -$ and m even. In this case, $G \not\leqslant \langle \mathrm{Inndiag}(T), \varphi^i \rangle$, so by [50, Tables 3.5.E and 3.5.G], any subgroup of G of type $\mathrm{GL}_m(q)$ is contained in a proper normal subgroup of G and is, therefore, not maximal.

The multiplicities follow quickly from Lemma 3.3.4 and Proposition 5.4.10. □

Lemma 5.4.13 *Assume that $m \neq 4$ and $(\eta, m) \neq (-, 5)$. A suitable power of y has type $I_{2m-2} \perp w$ where*

$$w = \begin{cases} (2)^-_{q_0} & \text{if } q_0 \text{ is not Mersenne} \\ -I_2 & \text{otherwise.} \end{cases}$$

Proof All of the types of elements that we discuss in this proof are defined over $\mathbb{F}_{q_0}$ and we omit the subscripts q_0 for simplicity of notation.

Case 1: q_0 Is Not Mersenne First assume that $\theta \in \{\varphi^i, \psi^i\}$. In this case, $y = y_1(\perp y_2) \perp (2)^-$, where $(2)^-$ has order $r = \Phi_2^*(q_0)$, and y_i has type $(d_i)^{\varepsilon_i}$ and order $r_i = \Phi^*_{\ell_i}(q_0)$, where we write $\ell_i = d_i/(d_i/2-1, 2)$ (we put the middle term in brackets to indicate that depending on m and η, the element might centralise a decomposition into either two or three subspaces). In particular, $\ell_i > 2$, so r and r_i are coprime. Consequently, a power of y has type $I_{2m-2} \perp (2)^-$, as required.

Next assume that $\theta \in \{\delta\varphi^i, \delta\psi^i\}$, so $y = {}^{\Delta}y_1(\perp {}^{\Delta}y_2) \perp {}^{\Delta}(2)^-$, where y_1 and y_2 are as in the previous case. Noting that $(2)^-$ has odd order, by Definition 5.3.13, we may raise y to a suitable power of 2 in order to obtain an element of type $y_1(\perp y_2) \perp (2)^-$, which reduces to the previous case.

Now assume that $\theta \in \{r_\square r_\boxtimes \varphi^i, r_\square r_\boxtimes \psi^i\}$, so y has type ${}^{\Sigma}y_1(\perp y_2) \perp (2)^-$. By Definition 5.3.15, we may again obtain an element of type $y_1(\perp y_2) \perp (2)^-$ by raising y to some suitable power of two, thus reducing this case to the first one.

Case 2: q_0 Is Mersenne In this case, an element of type $(2)^- = (2)^-_{q_0}$ has order q_0+1, which is a power of two, so we must be more careful when raising elements to even powers. However, note that elements of type $(2)^-$ and $(d)^\pm$ for $d > 2$ still have coprime order.

If $\theta \in \{r_{\Box}r_{\boxtimes}\varphi^i, r_{\Box}r_{\boxtimes}\psi^i\}$, then $y = y_1(\perp y_2) \perp (2)^-$, where y_i has type $(d_i)^{\varepsilon_i}$ and a power of y has type $I_{2m-2} \perp (2)^-$.

Now assume that $\theta \in \{\delta\varphi^i, \delta\psi^i\}$. For concreteness consider the case where $\eta = +$ and $m \equiv 0 \pmod 4$; the other cases are no harder to analyse. Here y has type ${}^\Delta(m)^- \perp {}^\Delta(m-2)^+ \perp {}^\Delta(2)^-$. Since q_0 is Mersenne, $q_0 \equiv 3 \pmod 4$ and consequently $(q_0^{m/2-1}+1)_2 = 2$, noting that $m/2-1$ is odd. Therefore, $y^{2(q_0-1)_2}$ has type $(m)^- \perp (m-2)^+ \perp x^2$, where x has type $(2)^-$. Now $|x| = q_0+1 \geqslant 4$ is a power of two, so $y^{\frac{1}{2}(q_0-1)_2(q_0+1)}$ has type $(m)^- \perp (m-2)^+ \perp -I_2$, a suitable (odd) power of which has type $I_{2m-2} \perp -I_2$.

Finally assume that $\theta = \{\varphi^i, \psi^i\}$. Then y has type ${}^\Sigma y_1(\perp y_2) \perp (2)^-$. Definition 5.3.15 informs us that y^2 has type $y_1(\perp y_2) \perp w$, where w has order $\frac{1}{2}(q_0+1) \geqslant 2$, so a power of y^2 has type $-I_2 \perp I_{2m-2}$. This completes the proof. □

Proposition 5.4.14 *Theorem 5.4.8 is true for primitive subgroups.*

Proof For now assume that $(\eta, m) \neq (-, 5)$. By construction, a suitable power of $t\theta$ is X-conjugate to y. By Lemma 5.4.13, fix a power $z = z' \perp I_{2m-2}$ of y, where $|z'|$ is a primitive prime divisor of $q_0^2 - 1$ if q_0 is not Mersenne and is 2 otherwise.

Now let $H \in \mathcal{M}(G, t\theta)$ be primitive. By Theorem 2.6.1, H is contained in one of the geometric families $\mathcal{C}_3, \dots, \mathcal{C}_8$ or is an almost simple irreducible group in the $\mathcal{S}$ family. We consider each family in turn.

Consider $\mathcal{C}_3$ subgroups. First suppose that H has type $\mathrm{O}^{\upsilon}_{2m/k}(q^k)$ for a prime divisor k of $2m$ and a sign $\upsilon \in \{\circ, \varepsilon\}$. Write $H \cap T = B.k$. From the definition of z, Lemma 5.3.23(ii) implies that $z \in B$. Moreover, since $\nu(z) = 2$, Lemma 5.3.23(i) implies that $k = 2$. Therefore, to verify the claim in Fig. 5.4, we can assume that m is even. In this case, a power of y has type $(2d)^+ \perp I_{2m-2d}$, where $d \in \{\frac{m}{2}, \frac{m-2}{2}, m-1\}$ is odd, which contradicts Corollary 5.3.25. Therefore, H does not have type $\mathrm{O}^{\upsilon}_{2m/k}(q^k)$ unless m is odd and $k = 2$.

Now suppose that H has type $\mathrm{GU}_m(q)$. These maximal subgroups only occur when $\varepsilon = +$ and m is even, or $\varepsilon = -$ and m is odd (see [50, Tables 3.5.E and 3.5.F]). Suppose that $\varepsilon = +$ but $\eta = -$ (and m is even). In this case a power of y has type $I_2 \perp (2m-2)^-$, but this is a contradiction to Corollary 5.3.25(ii)(a). Therefore, H has type $\mathrm{GU}_m(q)$ and $\varepsilon = \eta = (-)^m$.

Now let us turn to $\mathcal{C}_4$ subgroups. Suppose that H is the centraliser of a decomposition $V_1 \otimes V_2$ where $\dim V_1 \geqslant \dim V_2 > 1$. Since $z \in H$, we may write $z = z_1 \otimes z_2$. Since $\nu(z) = 2$, [55, Lemma 3.7] implies that $\nu(z_1) = 1$, $\nu(z_2) = 0$ and $\dim V_2 = 2$. Inspecting the conditions on $\dim V_1$ and $\dim V_2$ in [50, Tables 3.5.E and 3.5.F], this is impossible unless $\varepsilon = +$ and H has type $\mathrm{Sp}_m(q) \otimes \mathrm{Sp}_2(q)$. Since z is a semisimple element of prime order, $z_1 \in \mathrm{Sp}_m(q)$ is a semisimple element of prime order, but $\nu(z_1) = 1$, which is impossible. Therefore, $H \notin \mathcal{C}_4$.

If $H \in \mathcal{C}_5$, then H has type $\mathrm{O}^{\upsilon}_{2m}(q_1)$ where $q = q_1^k$ for a prime divisor k of f and a sign $\upsilon \in \{+, -\}$ such that $\upsilon^k = \varepsilon$.

The $\mathcal{C}_6$ family is empty since q is not prime.

We now treat $\mathcal{C}_7$ subgroups, which only arise when $\varepsilon = +$. Suppose that H is the stabiliser of a decomposition $U_1 \otimes U_2 \otimes \cdots \otimes U_k$ with $\dim U_i > 1$. Let $H_0 = H \cap \mathrm{PGL}(V)$ and write $H_0 = B.S_k$. Since z does not centralise a tensor product decomposition (see the discussion of $\mathcal{C}_4$ subgroups), $z \notin B$. Therefore, z cyclically permutes the k factors. However, z has prime order and exactly two nontrivial eigenvalues which contradicts the eigenvalue pattern required by [20, Lemma 5.7.2]. Therefore, $H \notin \mathcal{C}_7$.

The $\mathcal{C}_8$ family is empty.

Finally, consider the $\mathcal{S}$ family. Since $\nu(z) = 2$, $2m \geqslant 10$ and q is not prime, [41, Theorem 7.1] implies that no such subgroups arise.

It remains to assume that $(\eta, m) = (-, 5)$. To prove the result in this case, we simply note that y has type ${}^{a}(4)^{-} \perp {}^{c}(6)^{+}$, so a power of y has type $I_4 \perp (6)^{+}$, which, in light of Corollary 5.3.25, implies that y is not contained in subgroups of type $\mathrm{O}_5(q^2)$ or $\mathrm{GU}_5(q)$.

To complete the proof, we note that the stated upper bounds on the multiplicities of nonsubspace subgroups follow from Lemma 3.3.4 and Proposition 5.4.10. □

We have now proved Theorem 5.4.8 and are, consequently, in the position to prove Theorems 5.1 and 5.2 in Case I(a).

Proposition 5.4.15 *Let $G = \langle T, \theta \rangle \in \mathcal{A}$ with $T \neq \mathrm{P\Omega}_8^{\varepsilon}(q)$. In Case I(a), $u(G) \geqslant 2$ and as $q \to \infty$ we have $u(G) \to \infty$.*

Proof We apply the probabilistic method encapsulated by Lemma 2.1.1. Let $x \in G$ have prime order. We will obtain an upper bound on

$$P(x, t\theta) \leqslant \sum_{H \in \mathcal{M}(G, t\theta)} \mathrm{fpr}(x, G/H).$$

By Lemma 2.1.1 we need to show that $P(x, t\theta) < \frac{1}{2}$ and $P(x, t\theta) \to 0$ as $q \to \infty$.

Theorem 5.4.8 gives a superset of $\mathcal{M}(G, t\theta)$. Moreover, referring to Fig. 5.4, it is straightforward to show that

$$N = |C_{\mathrm{PDO}^{\eta}_{2m}(q_0)}(y)| \leqslant 2q_0^m.$$

For instance, if $\eta = -$ and m is even, then Lemmas 2.3.4 and 5.3.18 imply that

$$|C_{X_\sigma}(y)| \leqslant (q_0 + 1)(q_0^{m-1} - 1) \leqslant 2q_0^m.$$

The relevant fixed point ratios are given in Theorem 4.1.1 and Proposition 4.2.2, where we make use of the observation that $\nu(x) \geqslant 2$ for all $x \in G \cap \mathrm{PGO}^{\varepsilon}_{2m}(q)$.

Write $d(n)$ for the number of proper divisors of a number n.

First assume that $\eta = +$ and m is odd, or $\eta = -$ and m is even. Then

$$P(x, t\theta) \leqslant \frac{1}{q^2} + \frac{7}{q^{m-2}} + \frac{5}{q^{m-1}} + (2 + \log\log q + 2d(2m)) \cdot 2q_0^m \cdot \frac{3}{q^{2m-5}},$$

which proves $P(x, t\theta) \to 0$ as $q \to \infty$ and $P(x, t\theta) < \frac{1}{2}$ unless $(\eta, m, q) = (+, 5, 4)$. (Here we make use of the fact that when $\varepsilon = -$, we know that $2f/i$ is odd, so $i > 1$ and consequently $q \geqslant q_0^3$.)

In the exceptional case, $t\theta$ is not contained in a maximal parabolic subgroup, and we can discount subgroups of type $\mathrm{O}_{10}^-(2)$ since they do not contain elements of order $|y| = 51$. These observations, together with a refined bound on the centraliser $|C_{X_\sigma}(y)|$, give

$$P(x, t\theta) \leqslant \frac{1}{4^2} + \frac{3}{4^3} + \frac{1}{4^4} + (1+1) \cdot (2+1)(2^4+1) \cdot \frac{3}{4^5} < \frac{1}{2}.$$

Next assume that $\eta = +$ and m is even. Then

$$P(x, t\theta) \leqslant \frac{1}{q^2} + \frac{3}{q^{m/2-1}} + \frac{14}{q^{m-2}} + (1 + \log\log q + 2d(2m)) \cdot 2q_0^m \cdot \frac{3}{q^{2m-5}} + 8q_0^m \cdot \frac{3}{q^{2m-7}},$$

so $P(x, t\theta) \to 0$ as $q \to \infty$ and $P(x, t\theta) < \frac{1}{2}$, unless $(m, q) = (6, 4)$.

In this exceptional case, we will show that $t\theta$ is contained in no subgroups of type $\mathrm{GL}_6(4)$ or $\mathrm{GU}_6(4)$; omitting the corresponding term gives $P(x, \theta) < \frac{1}{2}$. The type of y is $(2)_2^- \perp (4)_2^- \perp (6)_2^+$. First suppose that y is contained in a subgroup H of type $\mathrm{GU}_6(4)$. Write $H \cap \mathrm{PGL}(V) = B.2$. A power y_1 of y has type $(2)_2^- \perp I_{10}$, whose order is 3. Therefore, $y_1 \in B$; however, $e = 2$, so this contradicts Corollary 5.3.25, so $t\theta$ is not contained in a $\mathrm{GU}_6(4)$ subgroup. Next suppose that y is contained in a subgroup H of type $\mathrm{GL}_6(4)$. Again we write $H \cap \mathrm{PGL}(V) = B.2$. A power y_2 of y has type $(4)_2^- \perp I_8$, whose order is 5. Therefore $y_2 \in B$. This implies that $y_2 = M \oplus M^{-\mathsf{T}}$. The four nontrivial eigenvalues of y_2 are $\lambda, \lambda^2, \lambda^{2^2}, \lambda^{2^3}$, where $|\lambda| = 5$. Without loss of generality, λ is an eigenvalue of M. On the one hand, λ^4 must be an eigenvalue of M, but, on the other hand, $\lambda^{-1} = \lambda^4$ is an eigenvalue of $M^{-\mathsf{T}}$, which is a contradiction. Therefore, $t\theta$ is not contained in a $\mathrm{GL}_6(4)$ subgroup.

Now assume that $\eta = -$ and $m \geqslant 7$ is odd. Then

$$P(x, t\theta) \leqslant \frac{1}{q^2} + \frac{2}{q^{(m-1)/2}} + \frac{11}{q^{m-3}} + \frac{1}{q^{m-5}} + (2 + \log\log q + 2d(2m)) \cdot 2q_0^m \cdot \frac{3}{q^{2m-5}} + 2q_0^m \cdot \frac{3}{q^{2m-7}} < \frac{1}{2}$$

and $P(x, t\theta) \to 0$ as $q \to \infty$.

Finally assume that $(\eta, m) = (-, 5)$. Then

$$P(x, t\theta) \leqslant \frac{1}{q^2} + \frac{8}{q^3} + \frac{4}{q^4} + (6 + \log\log q) \cdot 2q_0^5 \cdot \frac{3}{q^5} + 2q_0^5 \cdot \frac{3}{q^3},$$

which proves $P(x, t\theta) \to 0$ as $q \to \infty$ and $P(x, t\theta) < \frac{1}{2}$ unless $\varepsilon = +$ and $e = 2$. In this case, by arguing as above we can show that y is not contained in a subgroup of type $\mathrm{GL}_5(q)$ and omitting the corresponding term gives $P(x, t\theta) < \frac{1}{2}$ unless $q = 4$. Now assume that $q = 4$. Here we can discount subgroups of type $\mathrm{O}^+_{10}(2)$ since they do not contain elements of order $|y| = 35$ and, by Lemma 3.3.5, $t\theta$ is contained in at most $e^2 = 4$ subgroups of type $\mathrm{O}^-_{10}(2)$. Therefore,

$$P(x, t\theta) \leqslant \frac{1}{4^2} + \frac{8}{4^3} + \frac{4}{4^4} + (2 \cdot (2^2 + 1)(2^3 - 1) + 4) \cdot \frac{3}{4^5} < \frac{1}{2}.$$

and $P(x, t\theta) \to 0$ as $q \to \infty$. This completes the proof. □

In Case I(a), it remains to prove Theorems 5.1 and 5.2 with $T = \mathrm{P\Omega}^{\varepsilon}_8(q)$. Recall the element y was defined in Fig. 5.3 and Proposition 5.4.7 guarantees the existence of an element $t\theta \in T$ such that $F(t\theta) = y$.

Proposition 5.4.16 *Let $G = \langle T, \theta \rangle \in \mathcal{A}$ where $T = \mathrm{P\Omega}^{\varepsilon}_8(q)$. In Case I(a), $u(G) \geqslant 2$ and as $q \to \infty$ we have $u(G) \to \infty$.*

Proof We apply Lemma 2.1.1. First assume that $\eta = -$. Recall that e is even if $\varepsilon = +$ and e is odd if $\varepsilon = -$. In this case $y \in X_\sigma = \mathrm{PDO}^-_8(q_0)$ has type ${}^b(8)^-$, so $|C_{X_\sigma}(y)| \leqslant q_0^4 + 1$, by Lemma 5.3.18. Now y is not contained in any reducible subgroups of $\mathrm{PDO}^-_8(q_0)$, so by arguing as in the proof of Proposition 5.4.11, using Lemma 3.3.2, we deduce that $t\theta$ is not contained in any reducible subgroups of G. By [8, Tables 8.50-53], there are at most $M + \log\log q$ conjugacy classes of irreducible maximal subgroups of G where

$$M = \begin{cases} 6 \text{ if } \varepsilon = + \\ 2 \text{ if } \varepsilon = - \end{cases}$$

and G does not have any maximal subgroups of type $\mathrm{GL}^{\pm}_4(q)$. Therefore, from the bound in Proposition 4.2.2(ii), for all prime order $x \in G$ we have

$$P(x, t\theta) < (M + \log\log q)(q_0^4 + 1) \cdot \frac{3}{q^3} < \frac{1}{2}$$

and $P(x, t\theta) \to 0$ as $q \to \infty$, unless $e = 2$ and $f \in \{2, 4\}$ (so $\varepsilon = +$).

Now assume that $e = 2$ and $f \in \{2, 4\}$. Then a suitable power of y has type $(8)^-_{q_0} = A_1 \perp A_2$ centralising $\mathbb{F}_q^8 = U_1 \perp U_2$ where each U_i is a nondegenerate minus-type 4-space on which A_i acts irreducibly (see Lemma 5.3.17). Let us consider the possible imprimitive maximal overgroups of y of type $B{:}S_k$. Since the order of y is coprime to 24, $y \in B$, which implies that H has type $\mathrm{O}^-_4(q) \wr S_2$ and y^{12} (and hence $t\theta$) is contained in a unique conjugate of H. Now consider primitive overgroups. The subgroups of type $\mathrm{PSL}_3(q)$ and $\mathrm{O}^+_8(q^{1/2})$ contain no elements whose order is a primitive prime divisor of $(q^{1/2})^{16} - 1$, so y (and hence $t\theta$) is contained in no subgroups of these types, and by Lemma 3.3.5, $t\theta$ is contained in at most 4 subgroups of type $\mathrm{O}_8^-(q^{1/2})$. Therefore,

$$P(x,t\theta) < (1+3)\cdot\frac{3}{q^3} < \frac{1}{2}$$

and $P(x,t\theta) \to 0$ as $q \to \infty$.

Now assume that $\varepsilon = \eta = +$. Here $y \in \mathrm{PDO}_8^+(q_0)$ has type ${}^c(6)^- \perp {}^a(2)^-$, so $|C_{X_\sigma}(y)| \leqslant (q_0+1)(q_0^3+1)$. We now study $\mathcal{M}(G,t\theta)$, beginning with reducible subgroups. Since y is contained in a unique reducible maximal subgroup of $\mathrm{PDO}_8^+(q_0)$ (of type $\mathrm{O}_2^-(q_0)\times\mathrm{O}_6^-(q_0)$), by Lemma 3.3.2, we deduce that $t\theta$ is contained in a unique reducible maximal subgroup of G (of type $\mathrm{O}_2^\upsilon(q)\times\mathrm{O}_6^\upsilon(q)$ for some choice $\upsilon \in \{+,-\}$). Next note that G has six $\widetilde{G}$-classes of maximal imprimitive subgroups, exactly two of which have type $\mathrm{GL}_4(q)$. Finally consider primitive maximal subgroups. For each prime divisor k of f, there is one $\widetilde{G}$-class of subfield subgroups of type $\mathrm{O}_8^+(q^{1/k})$, and if f is even, then also one $\widetilde{G}$-class of $\mathrm{O}_8^-(q^{1/2})$ subgroups. There are at most 11 further $\widetilde{G}$-classes of maximal primitive subgroups, exactly two of which have type $\mathrm{GU}_4(q)$.

Let $x \in G$ have prime order. For now assume that $e \geqslant 3$. Then Theorem 4.1.1 and Proposition 4.2.2 imply that

$$P(x,t\theta) < \frac{4}{q^2}+\frac{1}{q^3}+(14+\log\log q)(q_0+1)(q_0^3+1)\frac{3}{q^{15/4}}+4(q_0+1)(q_0^3+1)\frac{2}{q^{12/5}} \to 0$$

$q \to \infty$ and $P(x,t\theta) < \frac{1}{2}$, unless $q = 2^3$.

Now assume that $q = 2^3$. Here $(t\theta)^3$ is X-conjugate to $y = (2)_2^- \perp (6)_2^-$. We will consider more carefully the maximal overgroups of $t\theta$.

We begin with imprimitive subgroups. Note that $y = (2)_2^- \perp (6)_2^- = A_0 \perp A_1 \perp A_2 \perp A_3$ centralising $\mathbb{F}_8^8 = U_0 \perp U_1 \perp U_2 \perp U_3$ where each U_i is a nondegenerate minus-type 2-space on which A_i acts irreducibly (indeed $|A_0| = 3$ and $|A_i| = 9$ if $i > 0$). This implies that $t\theta$ is not contained in any subgroups of types $\mathrm{GL}_4(8)$, $\mathrm{O}_4^-(8)\wr S_2$ or $\mathrm{O}_2^+(8)\wr S_4$ and is contained in at most 1 subgroup of type $\mathrm{O}_2^-(8)\wr S_4$ and at most 3 subgroups of type $\mathrm{O}_4^+(8)\wr S_2$.

We now turn to primitive subgroups. For subfield subgroups, by Lemma 3.3.5, $t\theta$ is contained in at most 9 subgroups of type $\mathrm{O}_8^+(2)$. For field extension subgroups, we claim that y is not contained in any subgroups of type $\mathrm{O}_4^+(8^2)$ and is contained in at most 16 subgroups of type $\mathrm{GU}_4(8)$. The first claim follows from Lemma 5.3.23 noting that $y^3 = [1,1,\lambda,\lambda,\lambda,\lambda^2,\lambda^2,\lambda^2]$ where $|\lambda| = 3$. For the second claim let H have type $\mathrm{GU}_4(8)$, write $H \cap T = B.2$ and let π be the field extension embedding. Now $y = [\lambda,\lambda^{-1},\mu,\mu^{-1},\mu^2,\mu^{-2},\mu^4,\mu^{-4}]$ where $|\lambda| = 3$ and $|\mu| = 9$. Let $b \in B$ satisfy $\pi(b) = y$. Then $y = [\lambda^{\varepsilon_1},\mu^{\varepsilon_2},\mu^{\varepsilon_3},\mu^{\varepsilon_4}]$ where $\varepsilon_i \in \{+,-\}$. Therefore, there are 16 possibilities for y up to B-conjugacy and consequently 8 possibilities up to H_0-conjugacy. Therefore, $|y^T \cap H_0| = 8|b^{H_0}|$. In addition, $|C_T(z)| = |\mathrm{GU}_1(8)||\mathrm{GU}_3(8)| = |C_{H_0}(b)|$, so by Lemma 2.1.2, we deduce that y is contained in 8 T-conjugates of H_0 and consequently 8 G-conjugates of H.

Since there are two G-classes of subgroups of type $\mathrm{GU}_4(8)$, we conclude that $t\theta$ is contained in at most 16 subgroups of G of type $\mathrm{GU}_4(8)$, as claimed. Finally, there are 5 further G-classes of maximal irreducible subgroups, so

$$P(x, t\theta) < \frac{4}{q^2} + \frac{1}{q^3} + (1+3+9+5(2+1)(2^3+1)) \cdot \frac{3}{8^{15/4}} + 16 \cdot \frac{2}{8^{12/5}} < \frac{1}{2}.$$

It remains to assume that $e = 2$. If $q_0 = 2$, then Proposition 5.2.19 implies that $u(G) \geqslant 2$, so we can assume that $q_0 \geqslant 3$. A power of y is $[\lambda, \lambda^{-1}] \perp [\mu, \mu^q, \mu^{q^2}, \mu^{-1}, \mu^{-q}, \mu^{-q^2}]$ with respect to $V = (U \oplus U^*) \perp (W \oplus W^*)$, where $|\lambda| > 2$ and $|\mu|$ is a primitive divisor of q^3-1. Therefore, by Lemma 2.3.3, $U \perp W$, $U \perp W^*$, $U^* \perp W$ and $U^* \perp W^*$ are the only totally singular subspaces stabilised by y, so y is contained in exactly two subgroups of G of type $\mathrm{GL}_4(q)$. Moreover, a power of y has type $I_2 \perp (6)^-_{q_0}$, so by Corollary 5.3.25, y is not contained in any subgroups of type $\mathrm{GU}_4(q)$. Therefore,

$$P(x, t\theta) < \frac{4}{q^2} + \frac{1}{q^3} + (14 + \log\log q)(q_0+1)(q_0^3+1) \cdot \frac{3}{q^{15/4}} + 2 \cdot \frac{2}{q^{12/5}} \to 0$$

as $q \to \infty$ and $P(x, t\theta) < \frac{1}{2}$, unless $q = 3^2$. Let $q = 3^2$. In this case, $|y|$ is divisible by 7, the unique primitive prime divisor of $3^6 - 1$, and the only types of irreducible maximal subgroup of G with order divisible by 7 are $\mathrm{GL}_4(9)$ (2 classes), $\mathrm{O}_8^+(3)$ (4 classes), $\mathrm{O}_8^-(3)$ and $\Omega_7(9)$. We know that $t\theta$ is contained in at most subgroups of type $\mathrm{GL}_4(9)$ and Lemma 3.3.5 implies that $t\theta$ is contained in at most 4 subgroups of type $\mathrm{O}_8^+(3)$. Thus we conclude that

$$P(x, t\theta) < \frac{4}{9^2} + \frac{1}{9^3} + (4 + 3(3+1)(3^3+1)) \cdot \frac{3}{9^{15/4}} + 2 \cdot \frac{2}{9^{12/5}} < \frac{1}{2}.$$

This completes the proof. □

5.4.2 *Case I(b)*

For Case I(b), we cannot select an element $t\theta \in T\theta$ by directly considering a Shintani map as we did in Case I(a). Indeed, this is precisely the reason for the distinction between Cases I(a) and I(b). Nevertheless, we can use Shintani descent indirectly to select appropriate elements in $T\theta$ via Lemma 3.4.1 (see Example 3.4.2).

Notation 5.4.17 Write $q = p^f$ where $f \geqslant 2$. Let $V = \mathbb{F}_q^{2m}$ with $m \geqslant 4$. Fix the simple algebraic group

$$X = \mathrm{PDO}_{2m}(\overline{\mathbb{F}}_p) = \begin{cases} \Omega_{2m}(\overline{\mathbb{F}}_2) & \text{if } p = 2 \\ \mathrm{PSO}_{2m}(\overline{\mathbb{F}}_p) & \text{if } p \text{ is odd.} \end{cases}$$

Fix the standard Frobenius endomorphism $\varphi = \varphi_{\mathcal{B}^+}$ of X, defined with respect to the standard basis $\mathcal{B}^+$, as $(a_{ij}) \mapsto (a_{ij}^p)$, modulo scalars.
With respect to the $\mathcal{B}^+$, write $V_E = \langle e_1, \dots, e_{m-1}\rangle$ and $V_F = \langle f_1, \dots, f_{m-1}\rangle$. With respect to the decomposition

$$V = (V_E \oplus V_F) \perp \langle e_m, f_m\rangle$$

recall that $r = I_{2m-2} \perp r^+$ and $\delta = \delta^+ = (\beta I_{m-1} \oplus I_{m-1}) \perp [\beta, 1]$, where, in the latter case q is odd and $\beta \in \mathbb{F}_q^\times$ has order $(q-1)_2$.
Fix

$$Z_1 = X^\circ_{(\langle e_m, f_m\rangle)} \cong \begin{cases} \Omega_{2m-2}(\overline{\mathbb{F}}_2) & \text{if } p = 2 \\ \mathrm{SO}_{2m-2}(\overline{\mathbb{F}}_p) & \text{if } p \text{ is odd} \end{cases}$$

and $Z_2 = (Z_1)_{(V_E \oplus V_F)} \cong \mathrm{GL}_{m-1}(\overline{\mathbb{F}}_p)$.

By Proposition 5.2.15, we assume that $\theta \in \mathrm{PGO}^+_{2m}(q)\varphi^i$ or $\theta \in \mathrm{PGO}^-_{2m}(q)\psi^i$.

Notation 5.4.17. (continued) Write $q = q_0^e$ and $e = f/i$.
Fix (σ, ρ, d, Z) as follows, where $\Delta = \delta\delta^{\sigma^{-1}}\delta^{\sigma^{-2}} \cdots \delta^{\sigma^{-(e-1)}}$

ε	θ	σ	ρ	d	Z
$+$	$r\varphi^i$	$r\varphi^i$	r	2	Z_1
	$\delta^- r\varphi^i$	$\delta r\varphi^i$	$r\Delta^{-1}$	$2(q_0-1)_2$	Z_2
$-$	ψ^i	φ^i	r	2	Z_1
	$\delta^-\psi^i$	$\delta\varphi^i$	$r\Delta^{-1}$	$2(q_0-1)_2$	Z_2

Remark 5.4.18 Let us comment on Notation 5.4.17.

(i) Note that Z_1 and Z_2 are connected φ-stable subgroups of X.
(ii) We have $Z_1 \leqslant C_X(r)$ since the map r is supported on $\langle e_m, f_m\rangle$.

(iii) If q is odd, then $Z_2 \leqslant C_{Z_1}(\delta|_{V_E \oplus V_F})$ since $\delta|_{V_E \oplus V_F}$ centralises the decomposition $V_E \oplus V_F$ and acts as a scalar on each summand.
(iv) The automorphisms ψ and δ^- of $\mathrm{P}\Omega^-_{2m}(q)$, where q is odd in the latter case, were introduced in (5.7) and Definition 5.2.8.
(v) Write $\widetilde{\sigma} = \sigma|_{X_{\rho\sigma^e}}$ and $\widetilde{\rho} = \rho|_{X_{\rho\sigma^e}}$. Observe that $X_{\rho\sigma^e}\widetilde{\sigma} = \mathrm{PDO}^{\varepsilon}_{2m}(q)\theta$, noting that when $\varepsilon = -$ we are making the usual identifications justified by the isomorphism $\Psi\colon X_{r\varphi^f} \to \mathrm{PDO}^-_{2m}(q)$ given in Lemma 2.5.4 (see Remark 5.4.2(iii)).

We now choose the elements for Case I(b) in the following proposition (see Remark 5.2.16 for an explanation of the statement and Fig. 5.5).

Proposition 5.4.19 *Let $T = \mathrm{P}\Omega^{\varepsilon}_{2m}(q)$ and let θ be an automorphism from Fig. 5.1 (in Case I(iii) or (v)). If y is the element in Fig. 5.5, then there exists $t \in T$ that centralises the decomposition $\langle e_1, \dots, f_{m-1}\rangle \perp \langle e_m, f_m\rangle$ such that $(t\theta)^e$ is X-conjugate to y. Moreover, if $H \leqslant G$, then the number of G-conjugates of H that contain $t\theta$ is at most $|C_{\mathrm{PDO}^{-\varepsilon}_{2m}(q_0)}(y^d)|$.*

Proof In each case, $(\rho\sigma^e)^d = \sigma^{ed}$. For instance, if $\varepsilon = +$ and $\theta = \delta r\varphi^i$, then

$$(\rho\sigma^e)^d = (r\Delta^{-1}\Delta(r\varphi^i)^e)^d = (\varphi^f)^{2(q_0-1)_2} = (\varphi^{2f})^{(q_0-1)_2}$$

and

$$\sigma^{ed} = (\delta r\varphi^i)^{ed} = (\Delta(r\varphi^i)^e)^d = (\Delta r\varphi^f)^d = (\Delta\Delta^r\varphi^{2f})^{(q_0-1)_2} = (\varphi^{2f})^{(q_0-1)_2}.$$

It is also easy to verify that $y\widetilde{\rho} \in Z_\sigma$. Therefore, Lemma 3.4.1 implies that there exists $g \in Z_{\sigma^e} \leqslant \mathrm{PSO}^{\varepsilon}_{2m}(q) \leqslant X_{\rho\sigma^e}$ such that $(g\widetilde{\sigma})^e$ is X-conjugate (indeed Z-conjugate) to y and if $H \leqslant G$, then the number of conjugates of H that contain $g\widetilde{\sigma}$ is at most $|C_{\mathrm{PDO}^{-\varepsilon}_{2m}(q_0)}(y^d)|$.

Generic case	
$m \pmod 4$	y
0 or 2	${}^a(2m-2)^+ \perp {}^a r^{\varepsilon}$
1	${}^a(m-3)^+ \perp {}^a(m+1)^+ \perp {}^a r^{\varepsilon}$
3	${}^a(m-5)^+ \perp {}^a(m+3)^+ \perp {}^a r^{\varepsilon}$

Specific cases		
m	θ	y
5 or 7	$r\varphi^i, \psi^i$	$(4)^- \perp (2m-6)^- \perp r^{\varepsilon}$
	$\delta r\varphi^i, \delta\psi^i$	$D^+_{2m-2} \perp {}^{\Delta}r^{\varepsilon}$

Note: we describe y by its type over $\mathbb{F}_{q_0}$ and D^+_{2m-2} is defined in Remark 5.4.20(ii)

Fig. 5.5 Case I(b): The element y for the automorphism θ

If $q^m \not\equiv \varepsilon \pmod 4$, then $\mathrm{PSO}^{\varepsilon}_{2m}(q) = T$ and $\widetilde{\sigma} = \theta$, so $g\widetilde{\sigma} \in T\theta$, as required (see (2.2)). Otherwise, $g \in \mathrm{PSO}^{\varepsilon}_{2m}(q) = T \cup T r_{\square} r_{\boxtimes}$, so we may choose $\theta \in \{\widetilde{\sigma}, r_{\square} r_{\boxtimes} \widetilde{\sigma}\}$ such that $g\widetilde{\sigma} \in T\theta$, which proves the claim, by Lemmas 5.2.6 and 5.2.14. □

Remark 5.4.20 We comment on the definition of $t\theta$ when $m \in \{5, 7\}$.

(i) Let $m = 5$ and let $\theta \in \{r\varphi^i, \psi^i\}$. By Fig. 5.5, $y = y_1 \perp y_2 \perp r^-$, centralising a decomposition $\mathbb{F}_{q_0}^{10} = U_1 \perp U_2 \perp U_3$, where y_1 and y_2 both have type ${}^{\Delta}(4)^-$. By Lemma 2.6.4, $\ell = \Phi_4^*(q_0) \geqslant 9$. Let Λ be the set of elements of order ℓ in $\mathbb{F}_{q_0}^{\times}$. Then $|\Lambda| \geqslant 8$, so we can, and will, assume that y_1 and y_2 have distinct sets of eigenvalues. This implies that U_1 and U_2 are nonisomorphic $\mathbb{F}_{q_0}\langle y \rangle$-modules.
(ii) Let q be odd and let $\theta \in \{\delta r\varphi^i, \delta\psi^i\}$. We define D^+_{2m-2} to be an element $\beta A \perp A^{-\mathsf{T}}$ where A is an irreducible element, whose order is a primitive prime divisor of $q_0^{m-1} - 1$. This is like, but not exactly the same as, an element of type ${}^{\Delta}(2m-2)^+$ (which does not exist when m is odd).

Continue to let T be the simple group $\mathrm{P}\Omega^{\varepsilon}_{2m}(q)$ and let θ be an automorphism from Fig. 5.1. Fix y from Fig. 5.5 and $t\theta \in G = \langle T, \theta \rangle$ from Proposition 5.4.19. The following result describes $\mathcal{M}(G, t\theta)$.

Theorem 5.4.21 *The maximal subgroups of G which contain $t\theta$ are listed in Figs. 5.6 and 5.7, where $m(H)$ is an upper bound on the multiplicity of the subgroups of type H in $\mathcal{M}(G, t\theta)$.*

Theorem 5.4.21 will be proved in parts. As before, write $\widetilde{G} = \langle X_{\sigma^e}, \widetilde{\sigma} \rangle$. We will make use of Proposition 5.4.10 in this section. We begin with reducible subgroups.

Proposition 5.4.22 *Theorem 5.4.21 is true for reducible subgroups.*

Proof Let us divide this proof into four parts.

Part 1: Setup Let $\mathcal{D}$ be the decomposition

$$V = V_1 \perp V_2 \quad \text{where} \quad V_1 = \langle e_1, \dots, f_{m-1} \rangle \quad \text{and} \quad V_2 = \langle e_m, f_m \rangle.$$

Observe that θ centralises $\mathcal{D}$, and write $\theta_i = \theta|_{V_i}$. By Proposition 5.4.19, t also centralises $\mathcal{D}$, so we may write $t\theta = t_1\theta_1 \perp t_2\theta_2$ with respect to $\mathcal{D}$. Let us also write $y = y_1 \perp {}^{a}r^{\varepsilon}$. We begin by studying the $\langle t_i\theta_i \rangle$-invariant subspaces of V_i.

Part 2: Subspaces of V_1 Let U_1 be a $\langle t_1\theta_1 \rangle$-invariant subspace of V_1. We will apply Lemma 3.4.1(ii)(b).

For the sake of brevity, let us assume that $m \geqslant 9$ and $m \equiv 1 \pmod 4$; the other cases are very similar and we comment on them below. In this case, the element y_1 has type ${}^{a}(m-3)^+_{q_0} \perp {}^{a}(m+1)^+_{q_0}$, where a is empty or Δ. Write $S = \langle e_1, \dots, f_{m-1} \rangle_{\mathbb{F}_{q_0}}$. Then y_1 centralises a decomposition $S = (S_1 \oplus S_2) \perp (S_3 \oplus S_4)$, where the S_i are pairwise nonisomorphic irreducible $\mathbb{F}_{q_0}\langle y_1 \rangle$-modules (here $\dim S_1 = \dim S_2 = \frac{m-3}{2}$ and $\dim S_3 = \dim S_4 = \frac{m+1}{2}$). Therefore, by Lemma 2.3.3, the only $\langle y_1 \rangle$-invariant subspaces of V are direct sums of S_1, S_2, S_3 and S_4.

	type of H	$m(H)$	conditions	
$\mathcal{C}_1$			$m \pmod 4$	q
	$\mathrm{O}_2^{v}(q) \times \mathrm{O}_{2m-2}^{\varepsilon v}(q)$	1		
	$\mathrm{Sp}_{2m-2}(q)$	1		even
	$\mathrm{O}_{2m-1}(q)$	2		odd
	P_{m-1}	2	even	
		4	odd	
	$\mathrm{O}_{m-3}^{v}(q) \times \mathrm{O}_{m+3}^{\varepsilon v}(q)$	1	1	
	$\mathrm{O}_{m-2} \times \mathrm{O}_{m+2}$	2	1	odd
	$\mathrm{O}_{m-1}^{v}(q) \times \mathrm{O}_{m+1}^{\varepsilon v}(q)$	1	1	
	$P_{(m-3)/2}$	2	1	
	$P_{(m+1)/2}$	2	1	
	$\mathrm{O}_{m-5}^{v}(q) \times \mathrm{O}_{m+5}^{\varepsilon v}(q)$	1	3	
	$\mathrm{O}_{m-4} \times \mathrm{O}_{m+4}$	2	3	odd
	$\mathrm{O}_{m-3}^{v}(q) \times \mathrm{O}_{m+3}^{\varepsilon v}(q)$	1	3	
	$P_{(m-5)/2}$	2	3	
	$P_{(m+3)/2}$	2	3	
$\mathcal{C}_2$	$\mathrm{O}_{2m/k}^{v}(q) \wr S_k$	N	$k \mid m,\ k > 1,\ v^k \in \varepsilon$	
	$\mathrm{O}_{2m/k}(q) \wr S_k$	N	$k \mid 2m,\ 2m/k > 1$ odd	
	$\mathrm{GL}_m(q)$	N	m odd, $\varepsilon = +$	
$\mathcal{C}_5$	$\mathrm{O}_{2m}^{v}(q^{1/k})$	N	$k \mid f$, k is prime, $v^k = \varepsilon$	

Note: $N = |C_{\mathrm{PDO}_{2m}^{-\varepsilon}(q_0)}(y^2)|$ and in $\mathcal{C}_1$ there is a unique choice of v

Fig. 5.6 Case I(b): Description of $\mathcal{M}(G, t\theta)$ for $m \notin \{5, 7\}$

We now proceed as in the proof of Proposition 5.4.11 (see that proof for more details), but we use Lemma 3.4.1(ii)(b) in place of Lemma 3.3.2. In particular, Lemma 3.4.1(ii)(b) establishes that the only possibilities for U_1 are direct sums of four pairwise nonisomorphic irreducible $\langle t\theta_1 \rangle$-invariant subspaces $U_{1,1}$, $U_{1,2}$, $U_{1,3}$ and $U_{1,4}$ (where $\dim U_{1,1} = \dim U_{1,2} = \frac{m-3}{2}$ and $\dim U_{1,3} = \dim U_{1,4} = \frac{m+1}{2}$). Moreover, we can deduce that these subspaces are totally singular but $U_{1,1} \oplus U_{1,2}$ and $U_{1,3} \oplus U_{1,4}$ are nondegenerate.

The other cases are very similar. In all cases U_1 is a direct sum of pairwise nonisomorphic irreducible $\mathbb{F}_q\langle y_1 \rangle$-submodules of dimension at least three. In particular, this implies that

$$\dim V_1 - \dim U_1 \notin \{1, 2\}. \tag{5.12}$$

Part 3: Subspaces of V_2 Next let U_2 be a $\langle t_2\theta_2 \rangle$-invariant subspace of V_2. Note that a power of $t_2\theta_2$ is ${}^{a}r^{\varepsilon}$. Therefore, if q is even, then Lemma 5.3.20 implies that there is at most one proper nonzero $\mathbb{F}_q\langle t_2\theta_2 \rangle$-invariant subspace of V_2. Similarly, if q is

type of H	$m(H)$	conditions		
$\mathcal{C}_1$		θ	m	q
$\mathrm{O}_2^{\upsilon}(q) \times \mathrm{O}_{2m-2}^{\varepsilon\upsilon}(q)$	1			
$\mathrm{O}_{2m-1}(q)$	2			odd
$\mathrm{Sp}_{2m-2}(q)$	1			even
P_{m-1}	2	$\delta r \varphi^i$ or $\delta \psi^i$		
$\mathrm{O}_4^{\upsilon}(q) \times \mathrm{O}_{2m-4}^{\varepsilon\upsilon}(q)$	1	$r\varphi^i$ or ψ^i		
$\mathrm{O}_6^{\upsilon}(q) \times \mathrm{O}_{2m-6}^{\varepsilon\upsilon}(q)$	1	$r\varphi^i$ or ψ^i		
$\mathrm{O}_5(q) \times \mathrm{O}_9(q)$	2	$r\varphi^i$ or ψ^i	7	odd
$\mathcal{C}_2$ $\mathrm{O}_2^{-}(q) \wr S_m$	N	e is even, $\varepsilon = -$ ($e = 2$ only if $m = 5$ and $\theta \in \{r\varphi^i, \psi^i\}$)		
$\mathrm{O}_m(q) \wr S_2$	N	q is odd		
$\mathcal{C}_3$ $\mathrm{O}_m(q^2)$	N	$\theta \in \{\delta r \varphi^i, \delta\psi^i\}$, e is odd		
$\mathrm{GU}_m(q)$	N	$\theta \in \{\delta r \varphi^i, \delta\psi^i\}$, e is odd, $\varepsilon = -$		
$\mathcal{C}_5$ $\mathrm{O}_{2m}^{\upsilon}(q^{1/k})$	N	$k \mid f$, k is prime, $\upsilon^k = \varepsilon$		

Note: $N = |C_{\mathrm{PDO}_{2m}^{-\varepsilon}(q_0)}(y^2)|$

Fig. 5.7 Case I(b): Description of $\mathcal{M}(G, t\theta)$ for $m \in \{5, 7\}$

odd, then Lemma 5.3.21 implies that there are at most two $\mathbb{F}_q\langle t_2\theta_2\rangle$-invariant proper nonzero subspaces of V_2.

Part 4: Subspaces of V Now let U be a $\langle t\theta\rangle$-invariant subspace of V. Let $\pi_i : U \to V_i$ be the projection map of U onto V_i. Then $U_i = \pi_i(U)$ is a $\langle t_i\theta_i\rangle$-invariant subspace of V_i.

Suppose that $U_2 \neq 0$ and $U_2 \nleqslant U$. We mimic the proof of Lemma 2.3.1. Let $W_i = U \cap U_i$. Let $u_1 \in U_1$ and let $u_2, v_2 \in U_2$ satisfy $u_1 + u_2 \in U$ and $u_1 + v_2 \in U$. Then $u_2 - v_2 \in U$, so $u_2 - v_2 \in W_2$. Therefore, $\{u_2 \in U_2 \mid u_1 + u_2 \in U\}$ is a coset of W_2, so there is a well-defined function $L \colon U_1 \to U_2/W_2$ where $L(u_1)$ is defined as $\{u_2 \in U_2 \mid u_1 + u_2 \in U\}$.

If $u_1, v_1 \in U_1$ and $u_2, v_2 \in U_2$ satisfy $u_1 + u_2 \in U$ and $v_1 + v_2 \in U$, then for all $\lambda \in \overline{\mathbb{F}}_q$ we have $(u_1 + u_2) + \lambda(v_1 + v_2) = (u_1 + \lambda v_1) + (u_2 + \lambda v_2)$, so

$$L(u_1 + \lambda v_1) = W_2 + (u_2 + \lambda v_2) = L(u_1) + \lambda L(v_1).$$

Therefore, L is linear.

For $u_1 \in U_1$, $L(u_1) = W_2$ if and only if $u_1 \in U$, so $\ker L = W_1$. Since $U_2 \nleqslant U$ we know that $U_2/W_2 \neq 0$. This implies that $\dim W_1 = \dim U_1 - \dim U_2/W_2 \in \{2m-3, 2m-4\}$. However, W_1 is a $\langle t_1\theta_1\rangle$-invariant subspace of V_1 and (5.12) implies that V_1 does not have a $\langle t_1\theta_1\rangle$-invariant subspace of dimension $2m-3$ or $2m-4$, so we have obtained a contradiction.

Therefore, either $U_2 = 0$ or $U_2 \leqslant U$. This implies that $U = U_1 \oplus U_2$, the possibilities for which follow from Parts 2 and 3. These exactly correspond to the subgroups given in Figs. 5.6 and 5.7. □

We now turn to irreducible subgroups.

Proposition 5.4.23 *Theorem 5.4.21 is true for irreducible subgroups.*

Proof

Case 1: $m \notin \{5, 7\}$ By construction, a suitable power of $t\theta$ is X-conjugate to y. We begin by demonstrating that we can fix a power z of y satisfying $|z| = 2$ and $1 \leqslant \nu(z) \leqslant 2$. If $(\varepsilon, \theta) \in \{(+, r\varphi^i), (-, \psi^i)\}$, then a power z of y has type $I_{2m-2} \perp r^\varepsilon$ and evidently $\nu(z) = 1$. Otherwise $(\varepsilon, \theta) \in \{(+, \delta r\varphi^i), (-, \delta\psi^i)\}$ and raising $y^{(q-1)_2}$ to a suitable power gives an element z of type $I_{2m-2} \perp -I_2$ and $\nu(z) = 2$.

Let $H \in \mathcal{M}(G, t\theta)$ be irreducible. We proceed as in the proof of Proposition 5.4.14, using Theorem 2.6.1. In particular, let us quickly handle the cases that are essentially identical to those in that previous proof. Observe that $\mathcal{C}_6$ and $\mathcal{C}_8$ are empty, z is not contained in an $\mathcal{S}$ family subgroup by [41, Theorem 7.1] and $\mathcal{C}_5$ subgroups have type $\mathrm{O}^{\upsilon}_{2m}(q_1)$ where $q = q_1^k$ for a prime k and a sign $\upsilon \in \{+, -\}$ such that $\upsilon^k = \varepsilon$.

The possible types of $\mathcal{C}_2$ subgroups are those given in Fig. 5.6 (see [50, Tables 3.5.E and 3.5.F]). The restriction on $\mathrm{GL}_m(q)$ subgroups arises for the reason given in the proof of Proposition 5.4.12 for $(\varepsilon, \eta) = (+, -)$.

Consider $\mathcal{C}_3$ subgroups. In this case, H is a field extension subgroup of type $\mathrm{O}^{\upsilon}_{2m/k}(q^k)$ or $\mathrm{GU}_m(q)$. If H has type $\mathrm{O}^{\upsilon}_{2m/k}(q^k)$, write $H \cap T = B.k$, and if H has type $\mathrm{GU}_m(q)$, write $H \cap T = B.b$ with $b = 2$ if m and q are even and $b = 1$ otherwise. Lemma 5.3.23(ii) implies that $z \in B$, and Lemma 5.3.23(i) implies that $k = 2$ since $\nu(z) \leqslant 2$. Now a power of y has type $(2d)^+ \perp I_{2m-2d}$ where $d \in \{m-1, \frac{m+1}{2}, \frac{m+3}{2}\}$ is odd and, in turn, a power of this is $w = w' \perp I_{2m-2d}$ where $|w|$ is a primitive prime divisor of $q_0^{2d} - 1$. Lemma 5.3.23(ii) implies that $w \in B$ and Corollary 5.3.25 implies that $w \notin B$ since d is odd, which is a contradiction. Therefore, $H \notin \mathcal{C}_3$.

For $\mathcal{C}_4$ subgroups, suppose that H is the centraliser of a decomposition $V_1 \otimes V_2$ where $\dim V_1 \geqslant \dim V_2 > 1$. Since $z \in H$, we may write $z = z_1 \otimes z_2$. If $\nu(z) = 1$, then we have a contradiction to [55, Lemma 3.7]. Otherwise $z = -I_2 \perp I_{2m-2}$ and we quickly deduce that $\varepsilon = +$, H has type $\mathrm{Sp}_m(q) \otimes \mathrm{Sp}_2(q)$ and $\nu(z_1) = 1$, which is not possible. Therefore, $H \notin \mathcal{C}_4$.

For $\mathcal{C}_7$ subgroups we may assume that $\varepsilon = +$. Suppose that $H = B.S_k$ is the stabiliser of a decomposition $U_1 \otimes U_2 \otimes \cdots \otimes U_k$. From the previous paragraph, $z \notin B$. However, [20, Lemma 5.7.2] implies that z does not cyclically permute the k factors, which is a contradiction. Therefore, $H \notin \mathcal{C}_7$.

To complete the proof when $m \notin \{5, 7\}$, we note that the stated upper bounds on the multiplicities of nonsubspace subgroups follow from Lemma 3.4.1(ii)(a) and Propositions 5.4.10 and 5.4.19.

Case 2: $m \in \{5, 7\}$ Let $H \in \mathcal{M}(G, t\theta)$ be irreducible. We proceed as in the previous case. In particular, note that a power z of y satisfies $\nu(z) \leqslant 2$, so by [41, Theorem 7.1] $H \notin \mathcal{S}$. Therefore, H is a geometric subgroup and by considering the possible types we see that it suffices to consider subgroups in $\mathcal{C}_2$, $\mathcal{C}_3$ and $\mathcal{C}_5$. The result is clear for $\mathcal{C}_5$ subgroups. Note also that the multiplicities, as usual, follow from Lemma 3.4.1(ii)(a) and Propositions 5.4.10 and 5.4.19.

First assume that H has type $\mathrm{O}_2^{\varepsilon}(q) \wr S_m$ stabilising a decomposition $\mathcal{D}$ of V into m nondegenerate 2-spaces. If e is odd, then a power of y is $I_2 \perp y'$ where y' centralises one of the following decompositions

$$-4 \perp -4, \quad -4 \perp -8, \quad +8, \quad +12, \quad +6 \perp +6$$

where $-d$ is an irreducible minus-type d-space and $+2d$ is the direct sum of a dual pair of two irreducible totally singular d-spaces. By [20, Lemma 5.2.6], y must centralise $\mathcal{D}$, which is a contradiction, since elements of these types act irreducibly on a space of dimension strictly greater than 2. Therefore, e is even. Now assume that $m = 7$ or $\theta \in \{\delta r_{\square}\varphi^i, \delta\psi^i\}$. If $e = 2$, then a power of y is $I_2 \perp y'$ where y' centralises one of the following decompositions

$$-4 \perp -4, \quad +4 \perp +4, \quad +6 \perp +6,$$

(with the notation above) and again we obtain a contradiction.

Next assume that $\varepsilon = +$ and H has type $\mathrm{GL}_m(q)$. Let H be the stabiliser of the decomposition $V = V_1 \oplus V_2$, where V_1 and V_2 are maximal totally singular subspaces of V. Record that e is odd since $\varepsilon = +$. If $\theta \in \{r\varphi^i, \psi^i\}$, then a power of y has type $I_2 \perp (4)_q^- \perp (2m-6)_q^-$, noting that $2m - 6 \in \{4, 8\}$, so y has odd order and does not stabilise a maximal totally singular subspace, which is a contradiction. Now assume that $\theta \in \{\delta r_{\square}\varphi^i, \delta\psi^i\}$. In this case, y has type ${}^{\Delta}r \perp {}^{\Delta}(2m-2)_{q_0}^+$. Therefore, y has type $M \perp (8)_q^+$ or $M \perp (6)_q^+ \perp (6)_q^+$, depending on whether m is 5 or 7, where M acts irreducibly on a 2-space (see Lemma 5.3.21). Now y^2 centralises the decomposition and we may assume that $U \subseteq V_1$, where U is a totally singular subspace of dimension 4 or 3 that is stabilised by y^2 and on which y^2 acts irreducibly. However, U is stabilised by y, so y stabilises V_1 and hence centralises the decomposition. However, since M is irreducible, y does not stabilise a maximal totally singular subspace, which is a contradiction. Therefore, $t\theta$ is not contained in a subgroup of type $\mathrm{GL}_m(q)$.

Now we may assume that H is a $\mathcal{C}_3$ subgroup. If $\theta \in \{r\varphi^i, \psi^i\}$, then a power z of y satisfies $\nu(z) = 1$, so y is not contained in H (see Lemma 5.3.23). Now assume $\theta \in \{\delta r\varphi^i, \delta\psi^i\}$ and H has type $\mathrm{O}_m(q^2)$ or $\mathrm{GU}_m(q)$. Note that $\varepsilon = -$ in the latter case (see [50, Table 3.5.E]). Since y has type ${}^{\Delta}(2m-2)_{q_0}^+ \perp {}^{\Delta}r_{q_0}$, y has exactly two eigenvalues, λ and $-\lambda$, of order $2(q_0+1)_2$. Lemma 5.3.23 implies that y arises from an element $g \in \Delta\mathrm{U}_m(q^2)$ or $\mathrm{GO}_m(q^2)$ with exactly one eigenvalue of order $2(q_0+1)_2$. Therefore, $\lambda^q = -\lambda$, so e is odd. This completes the proof. □

We have now proved Theorem 5.4.21, and we are now in a position to establish Theorems 5.1 and 5.2 in Case I(b). We consider two cases depending on whether $m \in \{5, 7\}$.

Proposition 5.4.24 *Let $G = \langle T, \theta \rangle \in \mathcal{A}$ where $T = \mathrm{P\Omega}_{2m}^{\varepsilon}(q)$ with $m \notin \{5, 7\}$. In Case I(b), $u(G) \geqslant 2$ and as $q \to \infty$ we have $u(G) \to \infty$.*

Proof Let $x \in G$ have prime order. Theorem 5.4.21 gives a superset of $\mathcal{M}(G, t\theta)$. Using the fixed point ratios from Theorem 4.1.1 and Proposition 4.2.2(i), we will prove that $P(x, t\theta) < \frac{1}{2}$ and $P(x, t\theta) \to 0$ as $q \to \infty$. For brevity, we will not explicitly note that $P(x, t\theta) \to 0$ as $q \to \infty$ separately in each case. Write $d(n)$ for the number of proper divisors of n.

Case 1: m Is Even In this case,

$$P(x, t\theta) \leqslant \frac{(2, q-1)}{q} + \frac{1}{q^2} + \frac{20}{q^{m-2}} + (1 + \log\log q + 2d(2m)) \cdot (q_0 + 1)(q_0^{m-1} - 1) \cdot \frac{2}{q^{m-2}},$$

so $P(x, t\theta) < \frac{1}{2}$ unless either $(m, q) \in \{(4, 8), (4, 27), (6, 8)\}$, or $e = f = 2$ and $m \leqslant 10$.

Consider the former case. The unique type of $\mathcal{C}_5$ subgroup is $\mathrm{O}^{\varepsilon}_{2m}(p)$. First assume $m = 6$ and $q = 8$, then a suitable power z of y has type $10_2^+ \perp I_2$, which has odd prime order and, over $\mathbb{F}_8$, acts irreducibly on a totally singular 5-space. This implies that z, and hence $t\theta$, is not contained in a $\mathcal{C}_2$ subgroup. Therefore, in this case,

$$P(x, t\theta) \leqslant \frac{1}{8} + \frac{1}{8^2} + \frac{20}{8^4} + (2+1)(2^5 - 1) \cdot \frac{2}{8^4} < \frac{1}{2}.$$

Next assume that $m = 4$ and $q \in \{8, 27\}$. For now assume that $\varepsilon = -$. The subgroups of type $\mathrm{O}_8^-(p)$ are the only nonsubspace subgroups containing $t\theta$. By Proposition 4.2.2, for subgroups H of this type we have $\mathrm{fpr}(x, G/H) < 3/q^3$ provided that $\nu(x) \neq 1$ and a direct calculation demonstrates that this bound also holds when $\nu(x) = 1$ in this case. With this, together with better bounds extracted from Theorem 4.1.1, we obtain

$$P(x, t\theta) \leqslant \frac{(2, q-1)}{q} + \frac{9}{q^2} + \frac{14}{q^3} + (q_0 + 1)(q_0^3 - 1) \cdot \frac{3}{q^3} < \frac{1}{2}.$$

If $\varepsilon = +$, then we must also take into account the $\mathcal{C}_2$ subgroups. Here $y = y_1 \perp y_2 \perp y_3 \perp y_4$, centralising $\mathbb{F}_q^8 = U_1 \perp U_2 \perp U_3 \perp U_4$, where y acts pairwise nonisomorphically on the U_i and each y_i acts on U_i by centralising the decomposition into two totally singular 1-spaces, acting nontrivially on both. In particular, y is not contained in any imprimitive subgroups of type $\mathrm{O}_2^-(q) \wr S_4$ or $\mathrm{O}_4^-(q) \wr S_2$ and is contained in at most one subgroup of type $\mathrm{O}_2^+(q) \wr S_4$ and at most three of type $\mathrm{O}_4^+(q) \wr S_2$. Therefore,

$$P(x, t\theta) \leqslant \frac{(2, q-1)}{q} + \frac{9}{q^2} + \frac{14}{q^3} + (q_0 + 1)(q_0^3 - 1) \cdot \frac{3}{q^3} + 4 \cdot \frac{2}{q^{12/5}} < \frac{1}{2}.$$

Now assume that $e = f = 2$ and $m \in \{4, 6, 8, 10\}$. Here $\varepsilon = -$ since e is even. Therefore, since $f = 2$, G has no $\mathcal{C}_5$ subgroups. We will now show that $t\theta$ is not contained in any $\mathcal{C}_2$ subgroups. Note that if q is odd, then $D(Q) = \boxtimes$ since

$q^m \equiv 1 \pmod 4$, so any $\mathcal{C}_2$ subgroup has type $\mathrm{O}^-_{2m/k}(q) \wr S_k$ where k is odd and $2m/k$ is even (see [50, Table 3.5.F]). If $m \in \{4, 8\}$, then no such subgroups arise. Now assume that $m \in \{6, 10\}$. The unique possible type of $\mathcal{C}_2$ subgroup is $\mathrm{O}^-_4(q) \wr S_{m/2}$. A power z of y has type $(2m-2)^+_{q_0} \perp I_2$, which acts irreducibly on two $(m-1)$-spaces over $\mathbb{F}_q$ since $e = 2$ and $m - 1$ is odd (see Lemma 5.3.17). By [20, Lemma 5.2.6], z must centralise a decomposition $U_1 \perp \cdots \perp U_{m/2}$ where $\dim U_i = 4$, which is impossible since y acts irreducibly on a totally singular subspace of dimension $m - 1 \geqslant 5$. Therefore, $t\theta$ is contained in no nonsubspace subgroups. Accordingly,

$$P(x, t\theta) \leqslant \frac{(2, q-1)}{q} + \frac{1}{q^2} + \frac{20}{q^{m-2}},$$

so $P(x, t\theta) < \frac{1}{2}$ unless $(m, q) = (4, 4)$. If $T = \Omega^-_8(4)$, Proposition 5.2.19 implies that $u(G) \geqslant 2$.

Case 2: m Is Odd If $m \equiv 1 \pmod 4$ and $m \geqslant 9$, then

$$P(x, t\theta) \leqslant \frac{(2, q-1)}{q} + \frac{1}{q^2} + \frac{2}{q^{(m-3)/2}} + \frac{6}{q^{(m-1)/2}} + \frac{56}{q^{m-3}}$$
$$+ (1 + \log\log q + 2d(2m) + q)(q_0 + 1)(q_0^{(m-3)/2} - 1)(q_0^{(m+1)/2} - 1)\frac{2}{q^{m-2}},$$

which proves that $P(x, t\theta) < \frac{1}{2}$ unless $(m, q) = (9, 4)$. In this exceptional case, $\varepsilon = -$ since e is even, so the only nonsubspace subgroup to arise has type $\mathrm{O}^-_2(q) \wr S_9$, so

$$P(x, t\theta) \leqslant \frac{1}{4} + \frac{1}{4^2} + \frac{2}{q^3} + \frac{6}{4^4} + \frac{56}{4^6} + (2+1)(2^3-1)(2^5-1) \cdot \frac{2}{4^7} < \frac{1}{2}.$$

If $m \equiv 3 \pmod 4$ and $m \geqslant 11$, then

$$P(x, t\theta) \leqslant \frac{(2, q-1)}{q} + \frac{1}{q^2} + \frac{2}{q^{(m-5)/2}} + \frac{6}{q^{(m+1)/2}} + \frac{56}{q^{m-5}}$$
$$+ (1 + \log\log q + 2d(2m) + q)(q_0 + 1)(q_0^{(m-5)/2} - 1)(q_0^{(m+3)/2} - 1)\frac{2}{q^{m-2}},$$

which proves that $P(x, t\theta) < \frac{1}{2}$ unless $(m, q) = (11, 4)$. In this case, as above, $\varepsilon = -$, the only type of nonsubspace subgroup to occur is $\mathrm{O}^-_2(q) \wr S_{11}$ and adjusting the bound accordingly demonstrates that $P(x, t\theta) < \frac{1}{2}$. This completes the proof. □

Proposition 5.4.25 *Let $G = \langle T, \theta \rangle \in \mathcal{A}$ where $T = \mathrm{P}\Omega^\varepsilon_{2m}(q)$ with $m \in \{5, 7\}$. In Case I(b), $u(G) \geqslant 2$ and as $q \to \infty$ we have $u(G) \to \infty$.*

Proof Let $x \in G$ have prime order. We proceed as in the previous proof. Theorem 5.4.21 gives a superset of $\mathcal{M}(G, t\theta)$, Theorem 4.1.1 and Proposition 4.2.2 give bounds on the associated fixed point ratios, and we will use this information to prove that $P(x, t\theta) < \frac{1}{2}$ and $P(x, \theta) \to 0$ as $q \to \infty$.

Case 1: $\theta \in \{\delta r\varphi^i, \delta\psi^i\}$ In this case q is odd and

$$P(x, t\theta) \leqslant \frac{2}{q} + \frac{1}{q^2} + \frac{10}{q^{m-2}} + \frac{10}{q^{m-1}} + (3+q+M) \cdot (q_0+1)(q_0^{m-1}-1) \cdot \frac{2}{q^{m-2}},$$

where M is the number of types of subfield subgroups. Notice that

$$M \leqslant \begin{cases} 0 & \text{if } f \text{ is a power of } 2 \\ 1 & \text{if } f \text{ is an odd prime power} \\ 1+\log\log q & \text{otherwise} \end{cases}$$

where in the first case $\varepsilon = -$ since e is even. With this bound on M we see that $P(x, t\theta) < \frac{1}{2}$ unless $(m, q) \in \{(7, 3^2), (7, 5^2)\}$, or $m = 5$ and either $f = e = 3$ or $e = 2$. If $(m, q) \in \{(7, 3^2), (7, 5^2)\}$, then $t\theta$ is contained in no $\mathcal{C}_3$ or $\mathcal{C}_5$ subgroups; adjusting the bound on $P(x, t\theta)$ accordingly proves that $P(x, t\theta) < \frac{1}{2}$.

Next assume that $m = 5$ and $f = e = 3$. If $\varepsilon = +$, then there are no subgroups of type $\mathrm{GU}_m(q)$, so

$$P(x, t\theta) \leqslant \frac{2}{q} + \frac{1}{q^2} + \frac{10}{q^3} + \frac{10}{q^4} + 4 \cdot (q_0+1)(q_0^4-1) \cdot \frac{2}{q^3} < \frac{1}{2}.$$

Therefore, assume that $\varepsilon = -$. If $x \notin \mathrm{PGL}(V)$ or $\nu(x) \geqslant 2$, then by Proposition 4.2.2(ii)

$$P(x, t\theta) \leqslant \frac{2}{q} + \frac{1}{q^2} + \frac{10}{q^3} + \frac{10}{q^4} + (4+q^2) \cdot (q_0+1)(q_0^4-1) \cdot \frac{3}{q^5} < \frac{1}{2},$$

while if $x \in \mathrm{PGL}(V)$ and $\nu(x) = 1$, then $\mathrm{fpr}(x, G/H) = 0$ for $\mathcal{C}_3$ subgroups H (see Lemma 5.3.23) and

$$P(x, t\theta) \leqslant \frac{2}{q} + \frac{1}{q^2} + \frac{10}{q^3} + \frac{10}{q^4} + 3 \cdot (q_0+1)(q_0^4-1) \cdot \frac{2}{q^3} < \frac{1}{2}.$$

Now assume that $m = 5$ and $e = 2$. In this case, the only type of nonsubspace subgroup to arise is $\mathrm{O}_5(q) \wr S_2$. We will now bound the number of subgroups of this type that contain $t\theta$. Note that a suitable power z of y has type $I_2 \perp (8)^+_{q_0}$, so $z = I_2 \perp A_1 \perp A_2$ centralising $\mathbb{F}_q^{10} = U_0 \perp U_1 \perp U_2$ where U_i is a plus-type 4-space and A_i centralises the decomposition of U_i into the direct sum of two dual totally singular 2-spaces on which it acts irreducibly. Then z stabilises

$q-1$ nondegenerate subspaces of U_0 and consequently stabilises exactly $2(q-1)$ nondegenerate 5-spaces of V (see Lemma 2.3.1). Therefore, z is contained in at most $q-1$ subgroups of type $\mathrm{O}_5(q) \wr S_2$, and thus

$$P(x, t\theta) < \frac{2}{q} + \frac{1}{q^2} + \frac{10}{q^3} + \frac{10}{q^4} + (1 + 3(q-1)) \cdot \frac{2}{q^3} < \frac{1}{2}.$$

Case 2: $\theta \in \{r\varphi^i, \psi^i\}$ If q is even, then

$$P(x, t\theta) \leqslant \frac{1}{q} + \frac{1}{q^2} + \frac{1}{q^4} + \frac{1}{q^{(m-1)/2}} + \frac{1}{q^{m-3}} + \frac{9}{q^{m-2}} + \frac{6}{q^{m-1}} + (2 + \log\log q) \cdot (q_0+1)(q_0^2+1)(q_0^{m-3}+1) \cdot \frac{2}{q^{m-2}},$$

and if q is odd, then

$$P(x, t\theta) \leqslant \frac{2}{q} + \frac{1}{q^2} + \frac{1}{q^4} + \frac{2}{q^5} + \frac{1}{q^{(m-1)/2}} + \frac{3}{q^{m-3}} + \frac{15}{q^{m-2}} + \frac{10}{q^{m-1}} + (3 + \log\log q) \cdot (q_0+1)(q_0^2+1)(q_0^{m-3}+1) \cdot \frac{2}{q^{m-2}}.$$

This proves that $P(x, t\theta) < \frac{1}{2}$ unless $(m, q) = (5, 8)$ or $e = 2$. If $(m, q) = (5, 8)$, then there is a unique type of subfield subgroup and $t\theta$ is not contained in a subgroup of type $\mathrm{O}_2^{\varepsilon}(q) \wr S_5$; adjusting the bound accordingly gives $P(x, t\theta) < \frac{1}{2}$.

Finally assume that $e = 2$. In this case $\varepsilon = -$ and no subfield subgroups arise. If $m = 7$, then $t\theta$ is not contained in a subgroup of type $\mathrm{O}_2^-(q) \wr S_7$, and adjusting the bound above accordingly, proves that $P(x, t\theta) < \frac{1}{2}$. If $m = 5$, then $y = (4)_{q_0}^- \perp (4)_{q_0}^- \perp r^- = A_1 \perp A_2 \perp A_3 \perp r^-$ centralising $\mathbb{F}_q^{10} = U_1 \perp U_2 \perp U_3 \perp U_4 \perp W$ where each U_i is a nondegenerate minus-type 2-space on which A_i acts irreducibly, so y is contained in a unique $\mathcal{C}_2$ subgroup of type $\mathrm{O}_2^{\varepsilon}(q) \wr S_5$. Therefore, if q is even, then

$$P(x, t\theta) \leqslant \frac{1}{q} + \frac{3}{q^2} + \frac{9}{q^3} + \frac{7}{q^4} + \frac{2}{q^3} < \frac{1}{2}.$$

Now assume that q is odd. Let H be a subgroup of type $\mathrm{O}_5(q) \wr S_2$ stabilising a decomposition $V_1 \perp V_2$. Now y^2 centralises the decomposition and we may assume that $U \subseteq V_1$, where U is one of the 2-spaces y^2 stabilises and on which y acts irreducibly. However, U is stabilised by y, so y stabilises V_1 and hence centralises the decomposition. By considering the number of choices for the stabilised 5-space containing the 1-eigenspace of y, we see that y is contained in at most $\binom{4}{2} = 6$

subgroups of type $O_5(q) \wr S_2$. Therefore,

$$P(x, t\theta) \leqslant \frac{2}{q} + \frac{3}{q^2} + \frac{1}{q^3} + \frac{4}{q^4} + \frac{15}{q^8} + 7 \cdot \frac{2}{q^{m-2}} < \frac{1}{2}.$$

This completes the proof. □

5.5 Case II: Linear Automorphisms

In this section, we begin with Case II. Accordingly, write $G = \langle T, \theta \rangle$ where $T = \mathrm{P\Omega}^{\varepsilon}_{2m}(q)$ for $m \geqslant 4$ and $\varepsilon \in \{+, -\}$ and where $\theta \in \mathrm{PGO}^{\varepsilon}_{2m}(q)$. Recall the cases

(a) $G \leqslant \mathrm{PDO}^{\varepsilon}_{2m}(q)$
(b) $G \nleqslant \mathrm{PDO}^{\varepsilon}_{2m}(q)$.

We will consider Cases II(a) and II(b) in Sects. 5.5.1 and 5.5.2, respectively.

5.5.1 *Case II(a)*

Let $m \geqslant 4$ and $\varepsilon \in \{+, -\}$. In this section, we focus on the groups $\mathrm{P\Omega}^{\varepsilon}_{2m}(q) \leqslant G \leqslant \mathrm{PDO}^{\varepsilon}_{2m}(q)$ and prove Theorems 5.1 and 5.2 in Case II(a). In [11], Breuer, Guralnick and Kantor proved that $u(T) \geqslant 2$. As they point out [11, p.447], their proofs, in fact, prove that $u(G) \geqslant 2$. The following result is motivated by this comment (see [21, Theorem 3.1] for a similar argument).

Proposition 5.5.1 *Let $G \in \mathcal{A}$. In Case II(a), $u(G) \geqslant 2$ and as $q \to \infty$ we have $u(G) \to \infty$.*

Proof If $G = T$, then the result follows from [11, Propositions 5.13–5.18] (the fact that $u(G) \to \infty$ as $q \to \infty$ is evident from the proofs). Now assume that q is odd and $\theta \in \{r_\square r_\boxtimes, \delta\}$. In the proofs of [11, Propositions 5.13–5.18], it is shown that for all prime order elements $x \in T$, we have that $P(x, s) < \frac{1}{2}$ and $P(x, s) \to 0$ as $q \to \infty$, for a suitable semisimple element $s \in T$. In each case, by Lemmas 5.3.3 and 5.3.4, there exists $g \in T\theta$ such that a suitable power of g is s. It is straightforward to verify that for all $x \in G$, we also have $P(x, g) < \frac{1}{2}$ and $P(x, g) \to 0$ and $q \to \infty$ and consequently $u(G) \geqslant 2$ and $u(G) \to \infty$ as $q \to \infty$. We give the details when $\varepsilon = +$ and $m \geqslant 7$ is odd with $\theta \in \{\delta, r_\square r_\boxtimes \delta\}$. The other cases are similar.

Assume that $\varepsilon = +$, $m \geqslant 7$ is odd and $\theta \in \{\delta, r_\square r_\boxtimes \delta\}$. Let $V = \mathbb{F}_q^{2m}$ be the natural module for T. By Lemma 5.3.4, there exists $x = x_1 \perp x_2 \in \mathrm{DO}^{+}_{2m}(q)$ centralising $V_1 \perp V_2$, where V_1 and V_2 are nondegenerate subspaces of dimensions $m-1$ and $m+1$, x_1 has order $(q-1)(q^{(m-1)/2}+1)$ acting irreducibly on V_1, x_2 has order $(q-1)(q^{(m+1)/2}+1)$ acting irreducibly on V_2 and $\tau(x_1) = \tau(x_2) = \alpha$ (where $\mathbb{F}_q^\times = \langle\alpha\rangle$). Since $\tau(x) = \alpha \notin (\mathbb{F}_q^\times)^2$, $g = xZ(\mathrm{DO}^+_{2m}(q)) \in \mathrm{PDO}^+_{2m}(q) \setminus \mathrm{PSO}^+_{2m}(q)$. Consequently, $g \in T\ddot{\delta} \cup Tr_\square r_\boxtimes \ddot{\delta}$, but $\ddot{\delta}$ and $\ddot{r}_\square \ddot{r}_\boxtimes \ddot{\delta}$ are $\mathrm{Out}(T)$-conjugate.

By Lemma 2.6.4, the order of g is divisible by a divisor $\ell > 2e+1$ of $\Phi_{m+1}(q)$. Therefore, by Theorem 2.6.3, all of the subgroups in $\mathcal{M}(G, g)$ are reducible, subfield or field extension subgroups. Since $m+1 > m$ and $(m+1, m) = 1$, the ℓ does not divide the order of any subfield or field extension subgroup of G. Therefore, we conclude that $\mathcal{M}(G, g)$ contains only reducible subgroups. Moreover, Lemma 2.3.3 implies that the only proper nonzero subspaces of V that are stabilised by g are V_1 and V_2. Consequently, $\mathcal{M}(G, g) = \{H\}$, where H has type $\mathrm{O}^-_{m-1}(q) \times \mathrm{O}^-_{m+1}(q)$.

Now Theorem 4.1.1 implies that for each prime order element $x \in G$ we have

$$P(x, g) \leqslant \mathrm{fpr}(x, G/\!/H) < \frac{1}{q^{(m+1)/2}} + \frac{2}{q^{m-2}} + \frac{2}{q^{m-1}} < \frac{1}{2}.$$

By Lemma 2.1.1 we conclude that $u(G) \geqslant 2$. Moreover, as $q \to \infty$ we have $P(x, g) \to 0$ and consequently $u(G) \to \infty$. □

5.5.2 *Case II(b)*

We now turn to Case II(b). By Proposition 5.2.15, we may assume that G is $\langle T, \theta\rangle$ where $T \in \mathcal{T}$ and $\theta \in \{r, \delta r\}$.

Recall the reflection r^ε defined in Definition 5.3.19, and if q is odd, the diagonal element δ^ε defined in Definitions 5.2.2 and 5.2.8. Unless there is ambiguity, we suppress the superscripts on r^ε and δ^ε. If q is odd, fix $\beta \in \mathbb{F}_q^\times$ of order $(q-1)_2$.

Remark 5.5.2 Proposition 5.2.19 implies that $u(G) \geqslant 2$ when G is one of

$$\mathrm{O}^\pm_8(2),\ \ \langle \mathrm{P}\Omega^\pm_8(3), r\rangle,\ \ \mathrm{O}^\pm_8(4),\ \ \mathrm{O}^\pm_{10}(2),\ \ \mathrm{O}^\pm_{12}(2). \tag{5.13}$$

Therefore, for the remainder of this section, we assume that G is not in (5.13).

We apply the probabilistic method, so we begin by selecting an element. For now assume that $T \neq \mathrm{P}\Omega^\pm_8(5)$. Let

$$y = \begin{cases} A \perp r & \text{if } \theta = r \\ {}^\Delta(2m-2)^- \perp {}^\Delta r & \text{if } \theta = \delta r, \end{cases}$$

where A has type $(2m-2)^-$, unless $q = 2$, in which case A has order $2^{m-1}+1$. If $T = \mathrm{P}\Omega^\pm_8(5)$, then let $y = A \perp r$ where A has order $(5^3+1)/2 = 63$ if $\theta = r$ and $(5^3+1)4 = 504$ if $\theta = \delta r$.

Proposition 5.5.3 *Let $G = \langle T, \theta \rangle$ for $T \in \mathcal{T}$ and $\theta \in \{r, \delta r\}$. Assume that G is not one of the groups in* (5.13).

(i) *If $\theta = r$, then $y \in Tr$.*
(ii) *If q is odd and y has type ${}^{\Delta}(2m-2)^{-} \perp {}^{\Delta}r$, then $y \in T\delta r$.*

Proof Part (i) is immediate since $I_2 \perp (2m-2)^{-} \in T$, by Lemma 5.3.11, and $I_2 \perp A$ is clearly in T when $q = 2$ and $(m, q) = (4, 5)$. Now consider part (ii), so q is odd. Let $x_1 \in \mathrm{DO}^{-}_{2m-2}(q)$ have type ${}^{\Delta}(2m-2)^{-}$, so $\tau(x_1) = \beta$ and $\det(x_1) = \beta^{m-1}$. Additionally, by Lemma 5.3.21(vi), $\tau({}^{\Delta}r^{-\varepsilon}) = \beta$ and $\det({}^{\Delta}r^{-\varepsilon}) = -\beta$. Therefore, the element $x = x_1 \perp {}^{\Delta}r^{-\varepsilon}$ has type ${}^{\Delta}(2m-2)^{-} \perp {}^{\Delta}r^{-\varepsilon}$ and satisfies $\tau(x) = \beta$ and $\det(x) = -\beta^{m}$. Let $y = xZ(\mathrm{DO}^{\varepsilon}_{2m}(q))$. Now $\tau(r) = 1$ and $\det(r) = -1$. Moreover, we saw in Remarks 5.2.3 and 5.2.9 that $\tau(\delta) = \beta$ and $\det(\delta) = \beta^{m}$. Therefore, $\tau(\delta r) = \beta$ and $\det(\delta r) = -\beta^{m}$. Consequently, $y \in \mathrm{PSO}^{\varepsilon}_{2m}(q)\delta r$, or in other words $y \in T\delta r \cup T r_{\square} r_{\boxtimes} \delta r$, but $\ddot{\delta}\ddot{r}$ and $\ddot{r}_{\square}\ddot{r}_{\boxtimes}\ddot{\delta}\ddot{r}$ are $\mathrm{Out}(T)$-conjugate. □

Theorem 5.5.4 *The maximal subgroups of G that contain y are listed in Fig. 5.8, where $m(H)$ is an upper bound on the multiplicity of the subgroups of type H in $\mathcal{M}(G, y)$.*

We will prove Theorem 5.5.4 in two parts, considering the reducible and irreducible maximal overgroups of y separately.

Proposition 5.5.5 *Theorem 5.5.4 is true for reducible subgroups.*

Proof First assume that q is odd and $\theta = \delta r$. Then y centralises an orthogonal decomposition $V = U \perp U^{\perp}$, where U is a nondegenerate 2-space. Moreover, y acts irreducibly on U and $U^{\perp}$ (see Lemma 5.3.21(iv)). Therefore, by Lemma 2.3.3, the only proper nonzero subspaces of V stabilised by y are U and $U^{\perp}$, so the only reducible maximal overgroup of y is one of type $\mathrm{O}^{-\varepsilon}_{2}(q) \times \mathrm{O}^{-}_{2m-2}(q)$.

Next assume that q is odd and $\theta = r$. In this case, the element y centralises a decomposition $V = U_1 \perp U_2 \perp (U_1 + U_2)^{\perp}$, where U_1 and U_2 are nondegenerate 1-spaces. Moreover, y acts irreducibly on $(U_1 + U_2)^{\perp}$, and acts as 1 and -1 on U_1 and U_2, respectively. Therefore, by Lemma 2.3.3, the only subspaces stabilised by y are direct sums of U_1, U_2 and $(U_1 + U_2)^{\perp}$. Consequently, the reducible maximal overgroups of y are two of type $\mathrm{O}_{2m-1}(q)$ (the stabilisers of U_1 and U_2) and one of type $\mathrm{O}^{-\varepsilon}_{2}(q) \times \mathrm{O}^{-}_{2m-2}(q)$ (the stabiliser of $U_1 + U_2$).

Finally assume that q is even and $\theta = r$. In this case, y centralises the decomposition $V = U \perp U^{\perp}$, where U is a nondegenerate 2-space. In this case, y acts irreducibly on $U^{\perp}$. However, y acts indecomposably on U and stabilises a

Fig. 5.8 Case II(b): Description of $\mathcal{M}(G, y)$

type of H	$m(H)$	conditions
$\mathrm{O}^{-\varepsilon}_{2}(q) \times \mathrm{O}^{-}_{2m-2}(q)$	1	
$\mathrm{O}_{2m-1}(q)$	2	q odd, $\theta = r$
$\mathrm{Sp}_{2m-2}(q)$	1	q even
$\mathrm{O}_{m}(q^2)$	4	q odd, m odd, $\theta = \delta r$

unique 1-dimensional (nonsingular) subspace W of U (see Lemma 5.3.20). Since there are no $\mathbb{F}_q\langle y\rangle$-homomorphisms between $U^\perp$ and any $\mathbb{F}_q\langle y\rangle$-subquotient of U, Corollary 2.3.2 implies that the only proper nonzero subspaces of V stabilised by y are W, U, $U^\perp$ and $U^\perp + W$. From this we deduce that the reducible maximal overgroups of y are one of type $\mathrm{Sp}_{2m-2}(q)$ (the stabiliser of W) and one of type $\mathrm{O}_2^{-\varepsilon}(q) \times \mathrm{O}_{2m-2}^-(q)$ (the stabiliser of U). □

Proposition 5.5.6 *Theorem 5.5.4 is true for irreducible subgroups.*

Proof Let $H \in \mathcal{M}(G, y)$ be an irreducible subgroup. Observe that we can fix a power z of y whose order is $\ell = \Phi^*_{2m-2}(q)$. By Lemma 2.6.4, if $q > 2$, then $\ell > 4m - 3$ unless $T = \mathrm{P}\Omega_8^+(5)$ (recall that we are assuming that G is not in (5.13)). Observe also that if $\theta = r$, then a power of y is r and $\nu(r) = 1$.

By Theorem 2.6.1 either H is a geometric subgroup contained in $\mathcal{C}_2 \cup \cdots \cup \mathcal{C}_8$ or H is an almost simple subgroup in $\mathcal{S}$. We begin by considering the geometric maximal overgroups H of y in G. Since $|y|$ is divisible by ℓ, which in turn is divisible by a primitive prime divisor of $q^{2m-2}-1$, the main theorem of [40] implies that the possibilities for H feature in [40, Examples 2.1–2.5]. Let us consider these possibilities in turn.

For orthogonal groups, Example 2.1 consists of subfield subgroups, none of which arise since for all proper divisors k of f, if $q_0 = p^k$, then ℓ does not divide

$$|\,\mathrm{O}^{\varepsilon}_{2m}(q_0)| = 2q_0^{m^2-m}(q_0^m - \varepsilon)\prod_{i=1}^{m-1}(q_0^{2i} - 1).$$

All subgroups in Example 2.2 are reducible.

Example 2.3 features the imprimitive subgroups of type $\mathrm{O}_1(q) \wr S_n$. For these we insist that $\varepsilon = +$, $q = p \geqslant 3$ and $\ell = 2m - 1$, which implies that $T = \mathrm{P}\Omega_8^+(5)$ (since otherwise $\ell > 4m - 3$). Now suppose that $T = \mathrm{P}\Omega_8^+(5)$ and H has type $\mathrm{O}_1(5) \wr S_8$. Then s^4 has order at least 63, but there are no elements of this order in $H \cap T = 2^7.A_8$. Therefore, no subgroups arise from Example 2.3.

The only possible field extension subgroup H in Example 2.4 is $\mathrm{O}^{\eta}_m(q^2)$ where $\eta = \varepsilon$ if m is even and $\eta = \circ$ if m is odd. If $\theta = r$, then a power of y is r and $\nu(r) = 1$, so y is not contained in such a subgroup, by [20, Lemma 5.3.2]. Now assume that $\theta = \delta r$. If m is even, then ℓ does not divide the order of H.

Therefore, if H is a field extension subgroup containing y, then q is odd, $\theta = \delta r$, m is odd and H has type $\mathrm{O}_m(q^2)$. We will now prove that, in this case, y is contained in four G-conjugates of H. Note that y is a semisimple element with eigenvalue multiset $\Lambda \cup \Lambda^q \cup \{\mu, \mu^q\}$, where $\Lambda = \{\lambda^{q^{2i}} \mid 0 \leqslant i \leqslant m-1\}$ for a scalar $\lambda \in \overline{\mathbb{F}}_p^\times$ of order $(q^m+1)_2(q-1)_2\ell$ and $\mu \in \overline{\mathbb{F}}_p^\times$ has order $2(q-1)_2$. Let $\pi : H \to G$ be the field extension embedding and write $H = B.\phi$, where ϕ is the field automorphism $\xi \mapsto \xi^q$. By [20, Lemma 5.3.2], if $\pi(\tilde{y}) = y$, then $\tilde{y}$ has one of the following eigenvalue sets

$$S_1 = \Lambda \cup \{\mu\}, \quad S_2 = \Lambda \cup \{\mu^q\}, \quad S_3 = \Lambda^q \cup \{\mu^q\}, \quad S_4 = \Lambda^q \cup \{\mu\}.$$

Let $\tilde{y}_i$ have eigenvalue set S_i. By [20, Propositions 3.4.3 and 3.5.4], $y^G \cap H = \bigcup_{i=1}^4 \tilde{y}_i^B$. Note that ϕ fuses $\tilde{y}_1^B$ with $\tilde{y}_3^B$ and fuses $\tilde{y}_2^B$ with $\tilde{y}_4^B$. Therefore, $y^G \cap H = \tilde{y}_1^H \cup \tilde{y}_2^H$. Since an element of type ${}^{\Delta}r^{\varepsilon}$ is self-centralising in $\mathrm{GO}_2^{\varepsilon}(q)$, Lemma 2.3.4 and [20, Appendix B] yield $|C_G(y)| = (q^{m-1}+1)(q-1)2 = 2|C_H(y)|$. Now Lemma 2.1.1 implies that the number of G-conjugates of H that contain y is

$$\frac{|y^G \cap H|}{|y^G|}\frac{|G|}{|H|} = \frac{2|C_G(y)|}{|C_H(y)|} = 4.$$

We now consider subgroups H contained in the $\mathcal{S}$ family. First assume that $\theta = r$. Suppose that q is prime and $\mathrm{soc}(H)$ is an alternating group in the representation afforded by the fully deleted permutation module. For now assume that $T \neq \mathrm{P\Omega}_8^{\pm}(5)$. If $q > 2$, then y has order 2ℓ where ℓ is either divisible by two primes at least $2m-1$ or a prime at least $4m-3$. If $q = 2$, then y has order $2(2^{m-1}+1)$, which is divisible by a prime at least $2m-1$. In both cases, S_{2m+2} does not contain an element of order $|y|$, so we conclude that $H \notin \mathcal{S}$. If $T = \mathrm{P\Omega}_8^{\pm}(5)$, then y^4 has order at least 63, but $H \cap T \cong A_{10}$ has no elements of order 63. Therefore H does not arise from the fully deleted permutation module. Therefore, since $\nu(y^{\ell}) = 1$, [41, Theorem 7.1] implies that $T = \mathrm{P\Omega}_8^+(q)$ with $q = p \geqslant 5$ and $\mathrm{soc}(H) = \mathrm{P\Omega}_8^+(2)$ (noting $G = \langle T, \theta\rangle$ does not have absolutely irreducible maximal subgroups of type $\Omega_7(q)$ or ${}^3D_4(q^{1/3})$, see [8, Table 8.50]) but again $\mathrm{soc}(H)$ contains no elements of order $|s^4|$.

Now assume that $\theta = \delta r$. If $T \neq \mathrm{P\Omega}_8^{\pm}(5)$, then Lemma 2.6.4 implies that $\ell > 4m-3$, so Theorem 2.6.3 eliminates all possibilities for H (see [39, Table 1], noting that $\langle \mathrm{P\Omega}_8^{\pm}(q), \delta r\rangle$ does not have any maximal absolutely irreducible subgroups of type $\Omega_7(q)$ or $\mathrm{PSU}_3(q)$, see [8, Tables 8.50 and 8.53]). If $T = \mathrm{P\Omega}_8^{\pm}(5)$, then $s^4 \in T$ has order at least 63 and no maximal subgroup in $\mathcal{S}$ contains an element of such an order. Therefore, no $\mathcal{S}$ family subgroups occur in this case either. □

Next we handle a special case in a more concrete fashion.

Proposition 5.5.7 *Let $G = \langle T, r\rangle$ with $m \geqslant 5$. Let $x_1, x_2 \in G$ have prime order and satisfy $\nu(x_1) = 1$ and $\nu(x_2) \leqslant 2$. Then there exists $g \in G$ such that $\langle x_1, y^g\rangle = \langle x_2, y^g\rangle = G$.*

Proof We prove the claim when q is odd; the case where q is even is similar. We work in terms of the bases $\mathcal{B}^{\varepsilon}$ in (2.7) and (2.8).

Let us fix three particular vectors. First let $t_1, t_{m-1} \in \langle e_1, f_1, e_{m-1}, f_{m-1}\rangle$ be nonsingular vectors such that $(e_i - f_i, t_i) = 0$ and $\langle e_i - f_i, t_i\rangle$ is a nondegenerate minus-type 2-space. Next let $t_2 \in \langle e_1, f_1, e_2 - f_2, e_{m-1}, f_{m-1}\rangle^{\perp}$ with the property that $\langle e_2 - f_2, t_2\rangle$ is a nondegenerate minus-type 2-space.

Recall that the element y has type $r^{\varepsilon} \perp (2m-2)^-$, centralising a decomposition $U_1 \perp U_2$. If $\varepsilon = -$, then we may assume that $r^+ = r_{e_1 - f_1}$ and

$$U_1 = \langle e_1, f_1\rangle \quad \text{and} \quad U_2 = \langle e_2, \dots, f_{m-1}, u_m, v_m\rangle.$$

If $\varepsilon = +$, then we may assume that $r^- = r_{e_1 - f_1}$ and

$$U_1 = \langle e_1 - f_1, t_1 \rangle \quad \text{and} \quad U_2 = \langle e_2, \dots, f_{m-2}, e_{m-1} - f_{m-1}, t_{m-1}, e_m, f_m \rangle.$$

Case 1: $\nu(x_2) = 1$ In this case, x_1 and x_2 are reflections in nonsingular vectors. If u_1 and u_2 are nonsingular vectors, then $r_{u_1} = r_{u_2}$ if and only if $\langle u_1 \rangle = \langle u_2 \rangle$. Therefore, it suffices to prove the claim for $x_1 = r_{u_1}$ and $x_2 = r_{u_2}$ for orbit representatives $(\langle u_1 \rangle, \langle u_2 \rangle)$ for the action of G on pairs of distinct nonsingular 1-spaces of V. We may assume that $u_1 = e_1 - f_1$. Now $V = \langle u_1 \rangle \perp \langle u_1 \rangle^{\perp}$ and $G_{\langle u_1 \rangle}$ acts transitively on the sets of nonzero vectors of a given norm in $\langle u_1 \rangle^{\perp}$. Therefore, we may assume that $u_2 = \xi u_1 + \eta(e_1 + f_1)$ or $u_2 = \xi u_1 + \eta e_3$ for scalars $\xi, \eta \in \mathbb{F}_q$. This amounts to the following two cases

(i) $u_2 = e_1 - \lambda f_1$ for $\lambda \in \mathbb{F}_q \setminus \{0, 1\}$
(ii) $u_2 = e_1 + f_1 + \lambda e_3$ for $\lambda \in \mathbb{F}_q^{\times}$

First assume that $\varepsilon = -$. Let z have type $r_v \perp (2m-2)^-$ centralising the decomposition $\langle v, w \rangle \perp \langle v, w \rangle^{\perp}$ where $v = e_1 + e_2 - f_2$ and $w = e_1 + e_2 + f_2$. Note that v is nonsingular and $\langle v, w \rangle$ is a nondegenerate plus-type 2-space. By Theorem 5.5.4, $\mathcal{M}(G, z) \subseteq \{G_{\langle v \rangle}, G_{\langle w \rangle}, G_{\langle v, w \rangle}\}$. Observe that $vx_1 = f_1 + e_2 + f_2$ and $wx_1 = f_1 + e_2 - f_2$, neither of which is contained in $\langle v, w \rangle$. Therefore, x_1 does not stabilise $\langle v \rangle$, $\langle w \rangle$ nor $\langle v, w \rangle$. Consequently, $\langle x_1, z \rangle = G$. Moreover, in the two possible cases above

(i) $vx_2 = \lambda f_1 + e_2 + f_2$ and $wx_2 = \lambda f_1 + e_2 - f_2$
(ii) $vx_2 = -f_1 + e_2 + f_2 - \lambda e_3$ and $wx_2 = -f_1 + e_2 + f_2 - \lambda e_3$

In both cases, vx_2 and wx_2 are not contained in $\langle v, w \rangle$, so, as above, $\langle x_2, z \rangle = G$. It remains to observe that since $Q(e_1 - f_1) = -2 = Q(e_1 + e_2 - f_2)$, there exists $g \in G$ such that $\langle e_1, f_1 \rangle g = \langle v, w \rangle$ and $(e_1 - f_1)g = e_1 + e_2 - f_2$. This implies that $r_{(e_1 - f_1)g} = r_v$ and $y^g = z$.

Now assume that $\varepsilon = +$. In this case, let z have type $r_v \perp (2m-2)^-$ centralising $\langle v, w \rangle \perp \langle v, w \rangle^{\perp}$ where $v = e_1 + e_2 - f_2$ and $w = e_1 + t_2$, noting that $\langle v, w \rangle$ is a nondegenerate minus-type 2-space. Arguing as in the previous case we see that $\langle x_1, z \rangle = \langle x_2, z \rangle = G$. Moreover, there exists $g \in G$ such that $\langle e_1 + f_1, t_1 \rangle g = \langle v, w \rangle$ and $(e_1 - f_1)g = e_1 + e_2 - f_2$, so $y^g = z$. This completes the proof in Case 1.

Case 2: $\nu(x_2) = 2$ and x_2 Is Semisimple In this case, x_1 is a reflection and x_2 centralises a decomposition $W \perp W^{\perp}$ where W is a nondegenerate 2-space. Moreover, if $|x_2| = 2$, then we may assume that $x_2 = -I_2 \perp I_{2m-2}$ and if $|x_2|$ is odd, then $x_2 = A \perp I_{2m-2}$ where A is irreducible. As in Case 1, it suffices to assume that $x_1 = r_u$ where $u = e_1 - f_1$ and consider orbit representatives W of the action of $G_{\langle u \rangle}$ on nondegenerate 2-subspaces of V. Considering that W is either plus- or minus-type, and by separating into the cases where

(i) $\langle u \rangle \leqslant W$ (ii) $W \leqslant \langle u \rangle^{\perp}$ (iii) $\langle u \rangle \not\leqslant W \not\leqslant \langle u \rangle^{\perp}$

we may assume that W is one of the following

(i) $W = \langle e_1, f_1\rangle$ or $W = \langle e_1 - f_1, t_1\rangle$
(ii) $W = \langle e_2, f_2\rangle$ or $W = \langle e_2 - f_2, t_2\rangle$
(iii) $W = \langle e_2 - f_2 + \lambda u, e_2 + f_2\rangle$ or $W = \langle e_2 - f_2 + \lambda u, t_2\rangle$ where $\lambda \in \mathbb{F}_q^{\times}$.

As in Case 1, let z be an element of type $r_v \perp (2m-2)^-$, centralising a decomposition $\langle v, w\rangle \perp \langle v, w\rangle^{\perp}$ where $v = e_1 + e_2 - f_2$. Moreover, let $w = e_1 + e_2 + f_2$ if $\varepsilon = -$ and $w = e_1 + t_2$ if $\varepsilon = +$. Note that $\langle v, w\rangle$ is a nondegenerate $(-\varepsilon)$-type 2-space. Consequently, we have $\langle x_1, z\rangle = G$. Since x_2 fixes $W^{\perp}$ pointwise and either negates or acts irreducibly on W, we see that $\langle x_2, z\rangle = G$ also.

Case 3: $\nu(x_2) = 2$ *and* x_2 *Is Unipotent* Here we need to consider the cases where x_2 has Jordan form $[J_2^2, J_1^{2m-4}]$ and $[J_3, J_1^{2m-3}]$. The latter case is very similar to Case 2, so we provide the details in the case where x_2 has Jordan form $[J_2^2, J_1^{2m-4}]$.

As before, x_1 is a reflection. In this case, x_2 centralises a decomposition $W \perp W^{\perp}$ where $W = W_1 \oplus W_2$ for totally singular 2-spaces W_1 and W_2. Moreover, x_2 acts trivially on $W^{\perp}$ and acts indecomposably on W_i stabilising a unique 1-space $\langle w_i\rangle \leqslant W_i$. As in the previous cases, it suffices to assume that $x_1 = r_u$ where $u = e_1 - f_1$ and consider orbits of the action of $G_{\langle u\rangle}$. In this way, we may assume that one of the following holds

(i) $W_1 = \langle e_1, e_2\rangle$ with $w_1 = e_1$ and $W_2 = \langle f_1, f_2\rangle$ with $w_2 = f_2$
(ii) $W_1 = \langle e_1, e_2\rangle$ with $w_1 = e_1 + e_2$ and $W_2 = \langle f_1, f_2\rangle$ with $w_2 = f_2$
(iii) $W_1 = \langle e_2, e_3\rangle$ with $w_1 = e_2$, and $W_2 = \langle f_2, f_3\rangle$ with $w_2 = f_3$
(iv) $W_1 = \langle e_2, e_1 + e_3\rangle$ with $w_1 = e_2$ and $W_2 = \langle f_2, f_3\rangle$ with $w_2 = f_3$
(v) $W_1 = \langle e_1 + e_2, e_3\rangle$ with $w_1 = e_1 + e_2$ and $W_2 = \langle f_2, f_3\rangle$ with $w_2 = f_3$

As in the previous cases, let z have type $r_v \perp (2m-2)^-$ centralising a decomposition $\langle v, w\rangle \perp \langle v, w\rangle^{\perp}$ where $v = e_1 + e_2 - f_2$, and let $w = e_1 + e_2 + f_2$ if $\varepsilon = -$ and $w = e_1 + t_2$ if $\varepsilon = +$. Consequently, we have $\langle x_1, z\rangle = G$. It is also easy to see that the action of x_2 on the decomposition $(W_1 \oplus W_2) \perp W^{\perp}$ ensures that x_2 stabilises none of $\langle v\rangle$, $\langle w\rangle$ and $\langle v, w\rangle$.

For example, consider case (i). Here

$$x_2 = \begin{pmatrix} 1 & 0 \\ 1 & 1 \end{pmatrix} \oplus \begin{pmatrix} 1 & -1 \\ 0 & 1 \end{pmatrix} \perp I_{2m-4}.$$

with respect to $(\langle e_1, e_2\rangle \oplus \langle f_1, f_2\rangle) \perp \langle e_1, f_1, e_2, f_2\rangle^{\perp}$. Therefore, x_2 fixes e_1 and f_2 and maps $e_2 \mapsto e_1 + e_2$ and $f_1 \mapsto f_1 - f_2$. Therefore, $vx_2, wx_2 \notin \langle v, w\rangle$. Therefore, we conclude that $\langle x_2, z\rangle = G$. □

Proposition 5.5.8 *Let* $G \in \mathcal{A}$. *In Case II(b),* $u(G) \geqslant 2$ *and as* $q \to \infty$ *we have* $u(G) \to \infty$.

Proof We will apply the probabilistic method encapsulated by Lemma 2.1.1. Theorem 5.5.4 gives the members of $\mathcal{M}(G, y)$. Let $x \in G$ have prime order. We now use fixed point ratio bounds from Sect. 4.1 to obtain an upper bound on $P(x, y)$.

If $\theta = \delta r$, then q is odd and

$$P(x, y) \leqslant \frac{1}{q^2} + \frac{1}{q^{m-1} - 1} + \frac{4}{q^{2m-3}} + \frac{1}{q^{2m-2}} + N_m \frac{2}{q^{m-2}} < \frac{1}{2}$$

where N_m is 4 if m is odd and 0 if m is even. In addition, $P(x, y) \to 0$ as $q \to \infty$.

From now on we may assume that $\theta = r$. By Remark 5.5.2, we may assume that G does not appear in (5.13). First assume that q is odd. For brevity, write

$$P_1(m, q) = \frac{1}{q^{m-1} - 1} + \frac{4}{q^m - 1} + \frac{4}{q^{2m-3}}.$$

In this case,

$$P(x, y) \leqslant 2q^{-1} + q^{-2} + q^{-(2m-2)} + 2q^{-(2m-1)} + P_1(m, q).$$

Now $P(x, y) \to 0$ as $q \to \infty$, and if $q > 3$, then $P(x, y) < \frac{1}{2}$. Now assume that $q = 3$ and therefore $m \geqslant 5$. Making use of the dependence on $\nu(x)$ in the fixed point ratio bounds in Proposition 4.1.3, we obtain

$$P(x, y) \leqslant \begin{cases} 2q^{-3} + q^{-6} + q^{-(2m-6)} + 2q^{-(2m-3)} + P_1(m, q) < 0.120 & \text{if } \nu(x) \geqslant 3 \\ 2q^{-2} + q^{-4} + q^{-(2m-4)} + 2q^{-(2m-2)} + P_1(m, q) < 0.268 & \text{if } \nu(x) = 2 \\ 2q^{-1} + q^{-2} + q^{-(2m-2)} + 2q^{-(2m-1)} + P_1(m, q) < 0.809 & \text{if } \nu(x) = 1 \end{cases}$$

Now let $x_1, x_2 \in G$ have prime order. If

$$P(x_1, y) + P(x_2, y) > 1$$

then we can assume that $\nu(x_1) = 1$ and $\nu(x_2) \leqslant 2$. In the latter case, Proposition 5.5.7 implies that there exists $g \in G$ such that $\langle x_1, y^g \rangle = \langle x_2, y^g \rangle = G$. Therefore, $u(G) \geqslant 2$.

Now assume that q is even. We proceed as when q is odd. In this case, write

$$P_2(m, q) = \frac{1}{q^{m-1} - 1} + \frac{2}{q^m - 1} + \frac{4}{q^{2m-3}}.$$

Here

$$P(x, y) \leqslant q^{-1} + q^{-2} + P_2(m, q).$$

Now $P(x, y) \to 0$ as $q \to \infty$, and if $q > 2$, then $P(x, y) < \frac{1}{2}$. Now assume that $q = 2$ and therefore $m \geqslant 7$. Now

$$P(x, y) \leqslant \begin{cases} q^{-3} + q^{-6} + P_2(m, q) < 0.175 \text{ if } \nu(x) \geqslant 3 \\ q^{-2} + q^{-4} + P_2(m, q) < 0.347 \text{ if } \nu(x) = 2 \\ q^{-1} + q^{-2} + P_2(m, q) < 0.784 \text{ if } \nu(x) = 1 \end{cases}$$

As above, for $x_1, x_2 \in G$ of prime order, if

$$P(x_1, y) + P(x_2, y) > 1$$

then we can assume that $\nu(x_1) = 1$ and $\nu(x_2) \leqslant 2$, in which case, Proposition 5.5.7 implies that there exists $g \in G$ such that $\langle x_1, y^g \rangle = \langle x_2, y^g \rangle = G$. Therefore, we conclude that $u(G) \geqslant 2$. □

5.6 Case III: Triality Automorphisms

This section sees the completion of the proofs of Theorems 5.1 and 5.2. Write $G = \langle T, \theta \rangle$ where $T = \mathrm{P\Omega}_8^+(q)$ and $\theta \in \mathrm{Aut}(T) \setminus \mathrm{P\Gamma O}_8^+(q)$. By Proposition 5.2.15, in Case III, it suffices to consider the following three cases

(a) $\theta = \tau\varphi^i$ where i is a proper divisor of f and 3 divides f/i
(b) $\theta = \tau\varphi^i$ where i is a proper divisor of f and 3 does not divide f/i
(c) $\theta = \tau$.

For Cases III(a) and III(b), we will apply Shintani descent and the application of Shintani descent will be very similar to that in Cases I(a) and I(b) respectively. In Case III(c), θ is a graph automorphism and the argument will be more reminiscent of Case II(b). It is worth noting that in all three cases $\nu(x) > 1$ for all $x \in G \cap \mathrm{PGO}_8^+(q)$. Cases III(a)–(c) will be considered in turn in Sects. 5.6.1–5.6.3, respectively.

5.6.1 *Case III(a)*

Write $q = p^f$ where $f \geqslant 2$. Let $V = \mathbb{F}_q^8$, and let $X = \mathrm{PDO}_8(\overline{\mathbb{F}}_p)$ be the simple algebraic group of adjoint type. Fix the standard Frobenius endomorphism $\varphi = \varphi_{\mathcal{B}^+}$ of X and the standard triality graph automorphism τ of X such that $C_X(\tau) = G_2(\overline{\mathbb{F}}_p)$.

Write $\sigma = \tau\varphi^i$ and $e = f/i$ and $q = q_0^e$. In Case III(a), we assume that 3 divides e. Let F be the Shintani map of (X, σ, e), so

$$F\colon \{(g\widetilde{\sigma})^{X_{\sigma^e}} \mid g \in X_{\sigma^e}\} \to \{x^{X_\sigma} \mid x \in X_\sigma\}.$$

Now $X_{\sigma^e} = \mathrm{Inndiag}(T)$, since 3 divides e, and $X_\sigma = T_0 = C_T(\varphi^i\tau) = {}^3D_4(q_0)$, the *Steinberg triality group*. Observe that the domain of F is, in fact, $\{(g\widetilde{\sigma})^{X_{\sigma^e}} \mid g \in T\}$ since every element of $\mathrm{Inndiag}(T)\widetilde{\sigma}$ is $\mathrm{Inndiag}(T)$-conjugate to an element of $T\widetilde{\sigma}$. Let $y \in T_0$ have order $q_0^4 - q_0^2 + 1$ and let $t \in T$ satisfy $F(t\theta) = y$.

Proposition 5.6.1 *Let $G = \langle T, \theta\rangle \in \mathcal{A}$. In Case III(a), $u(G) \geqslant 2$ and as $q \to \infty$ we have $u(G) \to \infty$.*

Proof The maximal subgroups of T_0 were determined in [49], see also [8, Table 8.50] for a convenient list. First note that the order of y does not divide the order of any parabolic subgroup of T_0. Therefore, by Lemma 3.3.2, we deduce that $t\theta$ is not contained in any parabolic subgroups of G (see Example 3.3.3). From [49, Table II], we see that $|C_{T_0}(y)| = q_0^4 - q_0^2 + 1$. By [8, Table 8.50], there are at most $10 + \log\log q$ classes of maximal nonparabolic subgroups of G. Note that all nonparabolic maximal subgroups of G are nonsubspace, see for example [17, Table 3.1]. Therefore, $u(G) \geqslant 2$ and $u(G) \to \infty$ as $q \to \infty$ since, noting that $e \geqslant 3$, for all prime order $x \in G$ we have

$$P(x, t\theta) < (10 + \log\log q)(q_0^4 - q_0^2 + 1)\cdot\frac{3}{q^{15/4}} < \frac{1}{2}. \qquad \square$$

5.6.2 *Case III(b)*

Write $q = p^f$ where $f \geqslant 2$. Let $X = \mathrm{PDO}_8(\overline{\mathbb{F}}_p)$. Fix the standard Frobenius endomorphism $\varphi = \varphi_{B^+}$ and the triality graph automorphism τ. Let Z be the centraliser $C_X(\tau) = G_2(\overline{\mathbb{F}}_p)$. Write $\sigma = \tau\varphi^i$ and $e = f/i$ and $q = q_0^e$. In Case III(b), we assume that 3 does not divide e.

Proposition 5.6.2 *Let $T = \mathrm{P}\Omega_8^+(q)$ and let $\theta = \tau\varphi^i$ where f/i is not divisible by* 3. *Let y have order 7 if $q_0 = 2$ and $q_0^2 - q_0 + 1$ if $q_0 > 2$. Then there exists $t \in T$ that commutes with τ such that $(t\theta)^e$ is X-conjugate to $y\tau^{-1}$. Moreover, if $H \leqslant G$, then the number of G-conjugates of H that contain $t\theta$ is at most $|C_{{}^3D_4(q_0)}(y^3)|$.*

Proof Since $(\tau\sigma^e)^3 = \varphi^{3f} = \sigma^{3e}$ and $y \in G_2(q_0) = Z_\sigma$, by Lemma 3.4.1, there exists $t \in Z_{\sigma^e} \leqslant \mathrm{P}\Omega_8^+(q) \leqslant X_{\tau\sigma^e}$ such that $(t\widetilde{\sigma})^e$ is X-conjugate (indeed Z-conjugate) to $y\tau^{-1}$ and if $H \leqslant G$, then the number of conjugates of H that contain $t\widetilde{\sigma}$ is at most $|C_{{}^3D_4(q_0)}(y^3)|$. $\square$

Lemma 5.6.3 *Assume that $q_0 > 2$. Let $z = y^3$, where $y \in G_2(q_0) \leqslant {}^3D_4(q_0)$ has order $q_0^2 - q_0 + 1$. Then $C_{{}^3D_4(q_0)}(z) = C_{q_0^2-q_0+1} \times C_{q_0^2-q_0+1}$.*

Proof We may assume that $z \in \mathrm{SU}_3(q_0) < G_2(q_0) < {}^3D_4(q_0)$, and consequently $z \in W < Z < X$, where $W = \mathrm{SL}_3(\overline{\mathbb{F}}_p)$, $Z = G_2(\overline{\mathbb{F}}_p)$ and $X = \mathrm{PSO}_8(\overline{\mathbb{F}}_p)$ are the corresponding algebraic groups. Let V and U be the natural modules for X and W, respectively, and observe that $V|_W = U \oplus U^* \oplus 0^2$, where 0 is the trivial module. By first considering the eigenvalues of z on U, and then on V via the given decomposition, we deduce that $C_X(z)^\circ$ is a maximal torus. In particular, this implies that z is a regular semisimple element of ${}^3D_4(q_0)$ and by inspecting [49, Table II], we deduce that $C_{{}^3D_4(q_0)}(z)$ is either $C_{q_0^2-q_0+1} \times C_{q_0^2-q_0+1}$ or $C_{q_0^3+1} \times C_{q_0+1}$. Finally, we observe that the $\mathrm{SU}_3(q_0)$ subgroup of $G_2(q_0)$ containing z is centralised in ${}^3D_4(q_0)$ by a torus of order $q_0^2 - q_0 + 1$ and this rules out the latter possibility. □

Proposition 5.6.4 *Let $G = \langle T, \theta \rangle \in \mathcal{A}$. In Case III(b), $u(G) \geqslant 2$ and as $q \to \infty$ we have $u(G) \to \infty$.*

Proof Write $z = y^3$. First assume that $q_0 > 2$. By Lemma 5.6.3, we have $|C_{{}^3D_4(q_0)}(z)| = (q_0^2 - q_0 + 1)^2$, and note that $|z|$ is divisible by a primitive prime divisor r of $q_0^6 - 1$. The only maximal parabolic subgroup of ${}^3D_4(q_0)$ with order divisible by r has type $H_0 = q^{1+8}{:}\,\mathrm{SL}_2(q^3).(q-1)$, but the maximal tori of $\mathrm{SL}_2(q^3)$ have order $q^3 \pm 1$, so there are no elements in H_0 with the appropriate centraliser in ${}^3D_4(q_0)$. Therefore, z is not contained in a maximal parabolic subgroup of ${}^3D_4(q_0)$. Now assume that $q_0 = 2$. In this case y and z have order 7 and it is straightforward to check that $|C_{{}^3D_4(2)}(z)| = 7^2$ and again that z is not contained in any parabolic subgroup of ${}^3D_4(2)$.

Suppose that $t\theta$ is contained in a parabolic subgroup of G. Then $t\theta$ is contained in a parabolic subgroup of $\mathrm{PDO}_8^+(q^3){:}\langle\theta\rangle$. Let F be the Shintani map of $(X, \sigma, 3e)$. Then Lemma 3.3.2 implies that $F(t\theta) = y^3 = z$ (see Lemma 3.2.5(ii)) is contained in a parabolic subgroup of ${}^3D_4(q_0)$, which we know is false. Thus we conclude that $t\theta$ is not contained in a parabolic subgroup of G.

Let M be 7^2 if $q_0 = 2$ and $(q_0^2-q_0+1)^2$ if $q_0 > 2$. There are at most $10+\log\log q$ classes of maximal nonparabolic subgroups of G, so for all prime order $x \in G$,

$$P(x, t\theta) < (10 + \log\log q) \cdot M \cdot \frac{3}{q^{15/4}} \to 0$$

as $q \to \infty$ and $P(x, \theta) < \frac{1}{2}$, unless $q = 4$. When $q = 4$, Proposition 5.2.19 implies that $u(G) \geqslant 2$. □

5.6.3 *Case III(c)*

Write $q = p^f$ where $f \geqslant 1$. Let τ be the triality graph automorphism and recall that $C_T(\tau) = G_2(q)$. If $q = 2$, then for $t \in G_2(2)$ of order 7, Proposition 5.2.19

gives $u(G) \geqslant 2$ witnessed by $t\tau$. From now on, assume that $q > 2$. In this case, let $t \in G_2(q)$ have order $q^2 - q + 1$.

Proposition 5.6.5 *Let $G = \langle T, \tau \rangle \in \mathcal{A}$. Then $u(G) \geqslant 2$ and as $q \to \infty$ we have $u(G) \to \infty$.*

Proof Let $z = (t\tau)^3 = t^3$. Since $q > 2$, the order $|z| = (q^2-q+1)/(q^2-q+1, 3)$ is divisible by some primitive prime divisor r of $q^6 - 1$. Let $H \in \mathcal{M}(G, t\tau)$. The possibilities for H are given in [8, Table 8.50]. The only G-classes of subgroup that have order divisible by r are those of type $\mathrm{O}_2^-(q) \times \mathrm{GU}_3(q)$ and $G_2(q)$, and if $q \equiv 2 \pmod 3$, then also an absolutely irreducible almost simple group with socle $\mathrm{PSU}_3(q)$.

First assume that H has type $\mathrm{O}_2^-(q) \times \mathrm{GU}_3(q)$. Let S be a maximal torus of $\mathrm{PSO}_8^+(q)$ that contains z. Since r divides $|S|$, we have $|S| = (q^3 + 1)(q + 1)$. Therefore, $z = A \perp B$ with respect to $\mathbb{F}_q^8 = U \perp W$, where U and W are nondegenerate minus-type subspaces of dimensions 6 and 2. Moreover, A has order dividing $q^3 + 1$ and B has order dividing $q + 1$. Since $(q + 1, q^2 - q + 1) = 1$, we deduce that $|A| = |z|$ and $|B| = 1$. Therefore, $z = A \perp I_2$, and since r divides $|z|$, Lemma 5.3.1 implies that A acts irreducibly on U. Write $H_0 = H \cap T$. Then $H_0 = K \cap K^{\tau} \cap K^{\tau^2}$, where K is the stabiliser in T of a nondegenerate minus-type 2-space of $\mathbb{F}_q^8$. Since z stabilises a unique such subspace, K is the unique T-conjugate of K containing z. Therefore, H is the unique G-conjugate of H containing z.

Next let $H = G_2(q) \times C_3$. By [48, Table II], any element of $G_2(q)$ with order $(q^2 - q + 1)/(q^2 - q + 1, 3)$ and centraliser in $G_2(q)$ of order $(q^2 - q + 1)/d$ for some $d \in \{1, (q^2 - q + 1, 3)\}$, in fact has a centraliser of order $q^2 - q + 1$. Let M be the number of H-classes that $z^G \cap H$ splits into. By consulting [28, 31], we see that there are at most $(q^2 - q)/6$ classes in $G_2(q)$ of elements whose centraliser has order $q^2 - q + 1$. In addition, by arguing as in [52, Lemma 4.5], $z^G \cap H$ splits into at most $|W(D_4)/W(G_2)| = 16$, classes, where $W(\Gamma)$ denotes the Weyl group of Γ. Therefore, $M \leqslant \min\{(q^2 - q)/6, 16\}$ and the number of G-conjugates of H that contain z is

$$\frac{|G|}{|H|}\frac{|z^G \cap H|}{|z^G|} \leqslant M\frac{|C_G(x)|}{|C_H(x)|} = M\frac{3(q^3+1)(q+1)}{3(q^2-q+1)} = M(q+1)^2.$$

Now assume that $q \equiv 2 \pmod 3$ and $H = \mathrm{PGU}_3(q) \times C_3$. The elements in $\mathrm{PGU}_3(q)$ of order $q^2 - q + 1$ act irreducibly on the natural module $\mathbb{F}_q^3$ and are self-centralising. Each $\mathrm{PGU}_3(q)$-class of such elements corresponds to an orbit under $\lambda \mapsto \lambda^q$ on the set Λ of elements of $\mathbb{F}_{q^6}^{\times}$ of order $q^2 - q + 1$. Since each of these orbits has size three, there are at most $(q^2 - q)/3$ such classes. Therefore, we the number of G-conjugates of H that contain z is

$$\frac{|G|}{|H|}\frac{|z^G \cap H|}{|z^G|} \leqslant \frac{q^2-q}{3}\frac{|C_G(x)|}{|C_H(x)|} = \frac{q^2-q}{3}\frac{3(q^3+1)(q+1)}{3(q^3+1)} = (q^3-q)/3.$$

Let $x \in G$ have prime order. By [41, Theorem 7.1], if $H \leqslant G$ has type $G_2(q)$ or $\mathrm{PGU}_3(q)$, then $x \in H \cap T$ only if $\nu(x) \geqslant 3$. Therefore, by Proposition 4.2.2,

$$P(x, t\tau) < \frac{2}{q^{12/5}} + M(q+1)^2 \frac{2}{q^{9/2}} + \delta_{2,(q \pmod 3)} \frac{q^3-q}{3} \frac{2}{q^{9/2}} < \frac{1}{2}$$

and $P(x, t\tau) \to 0$ and $q \to \infty$. □

Combining Propositions 5.4.15, 5.4.16, 5.4.24, and 5.4.25 in Case I, Propositions 5.5.1 and 5.5.8 in Case II, and Propositions 5.6.1, 5.6.4, and 5.6.5 in Case III, establishes Theorems 5.1 and 5.2.

Chapter 6
Unitary Groups

6.1 Introduction

In this final chapter we complete the proof of Theorems 4 and 5 by considering the unitary groups. Write $q = p^f$ and

$$\mathcal{T}_- = \{\mathrm{PSU}_n(q) \mid n \geqslant 3 \text{ and } (n,q) \notin (3,2)\} \tag{6.1}$$

$$\mathcal{A}_- = \{\langle T, \theta \rangle \mid T \in \mathcal{T} \text{ and } \theta \in \mathrm{Aut}(T)\}. \tag{6.2}$$

The subscript $-$ in this notation will be explained in (6.8) and (6.9). Note that we exclude the group $\mathrm{PSU}_3(2)$ from $\mathcal{T}$ since it is isomorphic to $3^2.Q_8$.

We now present the main theorems of this chapter.

Theorem 6.1 *If $G \in \mathcal{A}_-$, then $u(G) \geqslant 2$.*

Theorem 6.2 *Let (G_i) be a sequence of groups in $\mathcal{A}_-$ with $\mathrm{soc}(G_i) = \mathrm{PSU}_{n_i}(q_i)$. Then $u(G_i) \to \infty$ if $q_i \to \infty$.*

We partition our proof of Theorems 6.1 and 6.2 into two cases

(I) $\theta \in \mathrm{P\Gamma U}_n(q) \setminus \langle \mathrm{PGU}_n(q), \gamma \rangle$
(II) $\theta \in \langle \mathrm{PGU}_n(q), \gamma \rangle$

where γ is the standard involutory graph automorphism of $\mathrm{PGU}_n(q)$.

As in Chap. 5, we define two subcases of both Cases I and II

(a) $G \cap \langle \mathrm{PGU}_n(q), \gamma \rangle \leqslant \mathrm{PGU}_n(q)$
(b) $G \cap \langle \mathrm{PGU}_n(q), \gamma \rangle \nleqslant \mathrm{PGU}_n(q)$.

As we explain in Remark 6.2.8, one case in the proof of [21, Theorem 2] was omitted, and we take the opportunity to provide the proof of this case.

S. Harper, *The Spread of Almost Simple Classical Groups*, Lecture Notes in Mathematics 2286, https://doi.org/10.1007/978-3-030-74100-6_6

Theorem 6.3 *Let $T = \mathrm{PSL}_n(q)$, where $n \geqslant 4$ is even and $q = p^f$ is odd. Let $\theta \in \mathrm{PGL}_n(q)\gamma\varphi^i$, where φ is the standard field automorphism of T, γ is the standard graph automorphism of T and i is a proper divisor of f such that f/i is odd. Then $u(\langle T, \theta\rangle) \geqslant 2$ and $u(\langle T, \theta\rangle) \to \infty$ if $q \to \infty$.*

We proceed as in Chap. 5. Sections 6.2 and 6.3 establish general information about automorphisms and elements of linear and unitary groups. We then prove Theorems 6.1 and 6.2 in Cases I and II in Sects. 6.4 and 6.5, respectively, and Theorem 6.3 in Sect. 6.6.

6.2 Automorphisms

This section's aim is to determine the automorphisms $\theta \in \mathrm{Aut}(\mathrm{PSU}_n(q))$ we need to consider to prove Theorems 6.1 and 6.2. We simultaneously consider $\mathrm{PSL}_n^+(q) = \mathrm{PSL}_n(q)$ and $\mathrm{PSL}_n^-(q) = \mathrm{PSU}_n(q)$, where $q = p^f$ and $n \geqslant 3$ in both cases.

Fix $\varepsilon \in \{+, -\}$ and write $T = \mathrm{PSL}_n^\varepsilon(q)$. Let V be the natural module for T, so $V = \mathbb{F}_{q^d}^n$ where

$$d = \begin{cases} 1 \text{ if } \varepsilon = + \\ 2 \text{ if } \varepsilon = -. \end{cases} \tag{6.3}$$

Let $\mathcal{B} = (v_1, \ldots, v_n)$ be a basis for V, and assume that $\mathcal{B}$ is as in (2.10) if $\varepsilon = -$.

Recall from Definition 2.5.1, the standard Frobenius endomorphism defined as $\varphi = \varphi_{\mathcal{B}} \colon (x_{ij}) \mapsto (x_{ij}^p)$ and the standard graph automorphism $\gamma = \gamma_{\mathcal{B}} \colon x \to (x^{-\mathsf{T}})^J$, where J is the antidiagonal matrix with entries $1, -1, 1, -1, \ldots, (-1)^{n+1}$.

By [37, Tables 4.3.1 and 4.5.1],

$$C_{\mathrm{PGL}_n(\overline{\mathbb{F}}_p)}(\gamma) = \begin{cases} \mathrm{PGSp}_n(\overline{\mathbb{F}}_p) & \text{if } n \text{ is even} \\ \mathrm{PSO}_n(\overline{\mathbb{F}}_p) & \text{if } n \text{ is odd} \end{cases} \tag{6.4}$$

$$C_{\mathrm{PGL}_n^\varepsilon(q)}(\gamma) = \begin{cases} \mathrm{PGSp}_n(q) & \text{if } n \text{ is even} \\ \mathrm{PSO}_n(q) & \text{if } n \text{ is odd and } q \text{ is odd} \\ \mathrm{Sp}_{n-1}(q) & \text{if } n \text{ is odd and } q \text{ is even.} \end{cases} \tag{6.5}$$

Let $\alpha_\varepsilon \in \mathbb{F}_{q^2}^\times$ satisfy $|\alpha_\varepsilon| = q - \varepsilon$. We define one further element

Definition 6.2.1 Let $\hat{\delta}_\varepsilon \in \mathrm{GL}_n^\varepsilon(q)$ be the antidiagonal matrix with antidiagonal entries $(-1)^{\lfloor \frac{n}{2} \rfloor}\alpha_\varepsilon, 1, 1, \ldots, 1$ (from top-right to bottom-left), written with respect to $\mathcal{B}$. Let $\delta_\varepsilon \in \mathrm{PGL}_n^\varepsilon(q)$ be the image of $\hat{\delta}_\varepsilon$. If ε is understood, then we write $\delta = \delta_\varepsilon$.

Remark 6.2.2 It is easy to check that $\det(\hat{\delta}_\varepsilon) = \alpha_\varepsilon$ and $\delta_\varepsilon^\gamma = \delta_\varepsilon^{-1}$.

As in Chap. 5, for $g \in \mathrm{Aut}(T)$, we write $\ddot{g} = Tg \in \mathrm{Out}(T)$. From [50, Proposition 2.3.5], we have the outer automorphism groups

$$\mathrm{Out}(\mathrm{PSL}_n(q)) = \langle \ddot{\delta}, \ddot{\gamma}, \ddot{\varphi} \rangle = C_{(n,q-1)}{:}(C_2 \times C_f) \tag{6.6}$$

where $|\ddot{\delta}| = (n, q-1)$, $|\ddot{\gamma}| = 2$, $|\ddot{\varphi}| = f$, $\ddot{\delta}^{\ddot{\gamma}} = \ddot{\delta}^{-1}$ and $\ddot{\delta}^{\ddot{\varphi}} = \ddot{\delta}^p$, and

$$\mathrm{Out}(\mathrm{PSU}_n(q)) = \langle \ddot{\delta}, \ddot{\varphi} \rangle = C_{(n,q+1)}{:}C_{2f} \tag{6.7}$$

where $|\ddot{\delta}| = (n, q+1)$, $|\ddot{\varphi}| = 2f$ and $\ddot{\delta}^{\ddot{\varphi}} = \ddot{\delta}^p$.

We now present two similar lemmas that will be crucial to our case analysis.

Lemma 6.2.3 *Let $T = \mathrm{PSU}_n(q)$ and let i divide f. Then the following hold*

(i) *if $(n, q+1)$ is odd, then $(\ddot{\varphi}^i)^{\mathrm{Out}(T)} = \langle \ddot{\delta} \rangle \ddot{\varphi}^i$*
(ii) *if $(n, q+1)$ is even, then $(\ddot{\varphi}^i)^{\mathrm{Out}(T)} = \langle \ddot{\delta}^2 \rangle \ddot{\varphi}^i$ and $(\ddot{\delta}\ddot{\varphi}^i)^{\mathrm{Out}(T)} = \langle \ddot{\delta}^2 \rangle \ddot{\delta}\ddot{\varphi}^i$, so in particular $\langle \ddot{\delta} \rangle \ddot{\varphi}^i$ is the disjoint union $(\ddot{\varphi}^i)^{\mathrm{Out}(T)} \cup (\ddot{\delta}\ddot{\varphi}^i)^{\mathrm{Out}(T)}$.*

Proof Begin by observing that $(\ddot{\varphi}^i)^{\mathrm{Out}(T)} \subseteq \langle \ddot{\delta} \rangle \ddot{\varphi}^i$. Now let j be a divisor of $(n, q+1)$ and note that $\ddot{\delta}^j = (\ddot{\delta}^j)^{\ddot{\varphi}^i} = \ddot{\delta}^{jp^i}$ if and only if $(n, q+1)$ divides $(p^i - 1)j$. It is easy to see that $(q+1, p^i - 1) = (p-1, 2)$.

For (i), let us assume that $(n, q+1)$ is odd. Then $(n, q+1)$ and $p^i - 1$ are coprime, so $(n, q+1)$ divides $(p^i - 1)j$ if and only if $(n, q+1)$ divides j, that is, when $\ddot{\delta}^j = 1$. Consequently, $C_{\mathrm{Out}(T)}(\ddot{\varphi}^i) = \langle \ddot{\varphi} \rangle$, so $(\ddot{\varphi}^i)^{\mathrm{Out}(T)} = \langle \ddot{\delta} \rangle \ddot{\varphi}^i$.

For (ii), we now assume that $(n, q+1)$ is even. In this case, $((n, q+1), p^i - 1) = 2$, so $(n, q+1)$ divides $(p^i - 1)j$ if and only if $(n, q+1)/2$ divides j. Therefore, $C_{\mathrm{Out}(T)}(\ddot{\varphi}^i) = \langle \ddot{\varphi}, \ddot{\delta}^{(n,q+1)/2} \rangle$, so $\langle \ddot{\delta} \rangle \ddot{\varphi}^i$ must consist of exactly two $\mathrm{Out}(T)$-classes of equal size.

Let $h \in \langle \ddot{\delta}^2 \rangle \ddot{\varphi}^i$ and write $h = \ddot{\delta}^{2j} \ddot{\varphi}^i$. Note that $h^{\ddot{\varphi}^k} = \ddot{\delta}^{2jp^k} \ddot{\varphi}^i \in \langle \ddot{\delta}^2 \rangle \ddot{\varphi}^i$ and $h^{\ddot{\delta}^k} = \ddot{\delta}^{2j + k(p^i - 1)} \ddot{\varphi}^i \in \langle \ddot{\delta}^2 \rangle \ddot{\varphi}^i$. Therefore, $h^{\mathrm{Out}(T)} \in \langle \ddot{\delta}^2 \rangle \ddot{\varphi}^i$. This implies that $\langle \ddot{\delta}^2 \rangle \ddot{\varphi}^i$ is a union of conjugacy classes. However, since $|\ddot{\delta}| = (n, q+1)$ is even, $\langle \ddot{\delta} \rangle \ddot{\varphi}^i$ is the disjoint union of $\langle \ddot{\delta}^2 \rangle \ddot{\varphi}^i$ and $\langle \ddot{\delta}^2 \rangle \ddot{\delta} \ddot{\varphi}^i$, so these must be the two $\mathrm{Out}(T)$-classes in $\langle \ddot{\delta} \rangle \ddot{\varphi}^i$. Therefore, $(\ddot{\varphi}^i)^{\mathrm{Out}(T)} = \langle \ddot{\delta}^2 \rangle \ddot{\varphi}^i$ and $(\ddot{\delta}\ddot{\varphi}^i)^{\mathrm{Out}(T)} = \langle \ddot{\delta}^2 \rangle \ddot{\delta} \ddot{\varphi}^i$. □

Lemma 6.2.4 *Let $T = \mathrm{PSL}_n(q)$ and let i divide f and assume that f/i is odd. Then the following hold*

(i) *if $(n, q-1)$ is odd, then $(\ddot{\gamma}\ddot{\varphi}^i)^{\mathrm{Out}(T)} = \langle \ddot{\delta} \rangle \ddot{\gamma} \ddot{\varphi}^i$*
(ii) *if $(n, q-1)$ is even, then we have the classes $(\ddot{\gamma}\ddot{\varphi}^i)^{\mathrm{Out}(T)} = \langle \ddot{\delta}^2 \rangle \ddot{\gamma} \ddot{\varphi}^i$ and $(\ddot{\delta}\ddot{\gamma}\ddot{\varphi}^i)^{\mathrm{Out}(T)} = \langle \ddot{\delta}^2 \rangle \ddot{\delta} \ddot{\gamma} \ddot{\varphi}^i$, so in particular $\langle \ddot{\delta} \rangle \ddot{\gamma} \ddot{\varphi}^i$ is the disjoint union $(\ddot{\gamma}\ddot{\varphi}^i)^{\mathrm{Out}(T)} \cup (\ddot{\delta}\ddot{\gamma}\ddot{\varphi}^i)^{\mathrm{Out}(T)}$.*

Proof We argue just as in the proof of Lemma 6.2.3. First note $(\ddot{\gamma}\ddot{\varphi}^i)^{\mathrm{Out}(T)} \subseteq \langle \ddot{\delta} \rangle \ddot{\gamma} \ddot{\varphi}^i$. Now let j divide $(n, q-1)$ and note that $\ddot{\delta}^j = (\ddot{\delta}^j)^{\ddot{\gamma}\ddot{\varphi}^i} = \ddot{\delta}^{-jp^i}$ if and only if $(n, q-1)$ divides $(p^i + 1)j$. Note that $(q-1, p^i + 1) = (p-1, 2)$.

First assume that $(n, q-1)$ is odd. Then $(n, q-1)$ and p^i+1 are coprime, so $(n, q-1)$ divides $(p^i+1)j$ if and only if $(n, q-1)$ divides j, so $C_{\mathrm{Out}(T)}(\ddot{\gamma}\ddot{\varphi}^i) = \langle \ddot{\gamma}, \ddot{\varphi} \rangle$ and we conclude that $(\ddot{\gamma}\ddot{\varphi}^i)^{\mathrm{Out}(T)} = \langle \ddot{\delta} \rangle \ddot{\gamma}\ddot{\varphi}^i$.

Next assume that $(n, q-1)$ is even. Here $((n, q-1), p^i+1) = 2$, so $(n, q-1)$ divides $(p^i+1)j$ if and only if $(n, q-1)/2$ divides j, so $C_{\mathrm{Out}(T)}(\ddot{\varphi}^i) = \langle \ddot{\gamma}, \ddot{\varphi}, \ddot{\delta}^{(n,q-1)/2} \rangle$, and $\langle \ddot{\delta} \rangle \ddot{\gamma}\ddot{\varphi}^i$ consists of exactly two $\mathrm{Out}(T)$-classes.

By arguing as we did in the proof of Lemma 6.2.3, it is straightforward to show that $\langle \ddot{\delta}^2 \rangle \ddot{\gamma}\ddot{\varphi}^i$ is a union of conjugacy classes. Therefore, $(\ddot{\gamma}\ddot{\varphi}^i)^{\mathrm{Out}(T)} = \langle \ddot{\delta}^2 \rangle \ddot{\gamma}\ddot{\varphi}^i$ and $(\ddot{\delta}\ddot{\gamma}\ddot{\varphi}^i)^{\mathrm{Out}(T)} = \langle \ddot{\delta}^2 \rangle \ddot{\delta}\ddot{\gamma}\ddot{\varphi}^i$. □

Remark 6.2.5 Assume that $(n, q-\varepsilon)$ is even. Recall that $\delta \in \mathrm{PGL}^{\varepsilon}_n(q)$ lifts to an element $\hat{\delta} \in \mathrm{GL}^{\varepsilon}_n(q)$ of order $q-\varepsilon$. Now $|\delta| = (n, q-\varepsilon)$ and we define

$$\delta_2 = \delta^{\frac{q-\varepsilon}{(q-\varepsilon)_2}}.$$

(i) Note that $|\delta_2| = |\delta|_2 = (n, q-\varepsilon)_2$.
(ii) Since $\delta^{\gamma} = \delta^{-1}$ (see Remark 6.2.2), we have $\delta_2^{\gamma} = \delta_2^{-1}$, so $|\delta_2\gamma| = 2$.
(iii) As $\ddot{\delta}_2$ is not a square in $\langle \ddot{\delta} \rangle$, in view of Lemmas 6.2.3 and 6.2.4, the following hold
   (a) $(\ddot{\delta}_2\ddot{\varphi}^i)^{\mathrm{Out}(T)} = (\ddot{\delta}\ddot{\varphi}^i)^{\mathrm{Out}(T)}$ if $\varepsilon = -$ and i divides f
   (b) $(\ddot{\delta}_2\ddot{\gamma}\ddot{\varphi}^i)^{\mathrm{Out}(T)} = (\ddot{\delta}\ddot{\gamma}\ddot{\varphi}^i)^{\mathrm{Out}(T)}$ if $\varepsilon = +$ and i divides f with f/i odd.
(iv) As $\ddot{\gamma}$ and $\ddot{\delta}\ddot{\gamma}$ are not $\mathrm{Out}(T)$-conjugate $C_{\mathrm{PGL}_n(\overline{\mathbb{F}}_p)}(\delta_2\gamma) = \mathrm{PGO}_n(\overline{\mathbb{F}}_p)$ and $C_{\mathrm{PGL}_n(q)}(\delta_2\gamma) = \mathrm{PGO}^{\eta}_n(q)$ with $\eta = (-)^{\frac{n(q-\varepsilon)}{4}+1}$ (see [37, Table 4.5.1]).

We will now determine the almost simple linear and unitary groups it is sufficient to consider to prove our main theorems.

For a sign $\varepsilon \in \{+, -\}$, write

$$\mathcal{T}_{\varepsilon} = \{\mathrm{PSL}^{\varepsilon}_n(q) \mid n \geqslant 3 \text{ and } T \neq \mathrm{PSU}_3(2)\} \tag{6.8}$$

$$\mathcal{A}_{\varepsilon} = \{\langle T, \theta \rangle \mid T \in \mathcal{T} \text{ and } \theta \in \mathrm{Aut}(T)\}, \tag{6.9}$$

noting that this agrees with the definition of $\mathcal{T}_-$ and $\mathcal{A}_-$ in (6.1) and (6.2).

Proposition 6.2.6 *Let $G \in \mathcal{A}_{\varepsilon}$ with $\mathrm{soc}(G) = T = \mathrm{PSL}^{\varepsilon}_n(q)$. Then G is $\mathrm{Aut}(T)$-conjugate to $\langle T, \theta \rangle$ for one of the following*

(i) *θ in Row (1) of Fig. 6.1*
(ii) *θ in Row (2) of Fig. 6.1, if q is odd and n is even.*

Proof Write $G = \langle T, g \rangle$ where $g \in \mathrm{Aut}(T)$. We first consider $T = \mathrm{PSU}_n(q)$. From the description of $\mathrm{Out}(T)$, we see that we may write $\ddot{g} = \ddot{\delta}^{\ell}\ddot{\varphi}^i$ where $0 \leqslant \ell < (n, q+1)$ and $0 \leqslant i < 2f$. By Lemma 5.2.1 we may assume that $i = 0$ or i divides $2f$. If $i = 0$, then $\ddot{g} = \ddot{\delta}^{\ell}$ and we are in Case II(a), and if $i > 0$ and $2f/i$ is odd, then we are in Case I(a). Therefore, from now on, we will assume that i divides f.

Fig. 6.1 The relevant automorphisms θ when $T = \mathrm{PSL}_n^{\varepsilon}(q)$

				I(a)	I(b)	II(a)	II(b)	
ε	$+$	$+$	$+$	$-$	$-$			
θ	$\delta^{\ell}\varphi^i$	$\delta^{\ell}\gamma\varphi^i$	$\gamma\varphi^i$	$\delta^{\ell}\varphi^i$	φ^i	δ^{ℓ}	γ	(1)
			$\delta_2\gamma\varphi^i$		$\delta_2\varphi^i$		$\delta_2\gamma$	(2)
df/i	any	even	odd	odd	even			

Note: i is a proper divisor of df and $0 \leqslant \ell \leqslant (n, q-\varepsilon)$

First assume that n is odd or q is even, so $(n, q+1)$ is odd. By Lemma 6.2.3(i), $\ddot{g} = \ddot{\delta}^{\ell}\ddot{\varphi}^i$ is $\mathrm{Out}(T)$-conjugate to $\ddot{\varphi}^i$. If $i < f$, then we are in Case I(b). If $i = f$, then noting that $\ddot{g} = \ddot{\varphi}^f = \ddot{\gamma}$, we see that we are in Case II(b).

Now assume that n is even and q is odd, so $(n, q+1)$ is even. In this case, Lemma 6.2.3(ii) implies that $\ddot{g}$ is $\mathrm{Out}(T)$-conjugate to either $\ddot{\varphi}^i$ or $\ddot{\delta}_2\ddot{\varphi}^i$, where $\delta_2 = \delta^{\frac{q+1}{(q+1)_2}}$. If $i < f$, then we are in Case I(b). Since $\ddot{\varphi}^f = \ddot{\gamma}$ and $\ddot{\delta}_2\ddot{\varphi}^f = \ddot{\delta}_2\ddot{\gamma}$, if $i = f$, then we deduce that we are in Case II(b). This completes the proof for $T = \mathrm{PSU}_n(q)$.

It remains to consider $T = \mathrm{PSL}_n(q)$. As usual, we may assume that $\ddot{g} = \ddot{h}\ddot{\varphi}^i$ where h is a product of diagonal and graph automorphisms and where either $i = 0$ or i divides f. We claim that there is an automorphism θ in the statement such that $\ddot{g} = \ddot{h}\ddot{\varphi}^i$ is $\mathrm{Out}(T)$-conjugate to $\ddot{\theta}$. This is clear if h is diagonal or f/i is even. Therefore, assume that $\ddot{h} = \ddot{\delta}^{\ell}\ddot{\gamma}$ and f/i is odd. Then Lemma 6.2.4 implies that $\ddot{g}$ is $\mathrm{Out}(T)$-conjugate to $\ddot{\gamma}\ddot{\varphi}^i$ or, if $(n, q-1)$ is even, $\ddot{\delta}_2\ddot{\gamma}\ddot{\varphi}^i$ where $\delta_2 = \delta^{\frac{q-1}{(q-1)_2}}$, as required. □

Remark 6.2.7 It is easy to check that the automorphisms θ given in Proposition 6.2.6 when $\varepsilon = +$ and $f/i \geqslant 1$ is odd or $\varepsilon = -$ and $2f/i \geqslant 2$ is even are pairwise not $\mathrm{Out}(T)$-conjugate, and the $\mathrm{Out}(T)$-classes in the remaining cases are as follows

ε	g	conditions	classes in $\mathrm{Outdiag}(T)g$
$+$	$\ddot{\varphi}^i$	none	$\{\ddot{\delta}^j, \ddot{\delta}^{-j}, \dots, \ddot{\delta}^{jp^{f-1}}, \ddot{\delta}^{-jp^{f-1}}\}\langle\ddot{\delta}^{p^i-1}\rangle g$
	$\ddot{\gamma}\ddot{\varphi}^i$	f/i even	$\{\ddot{\delta}^j, \ddot{\delta}^{-j}, \dots, \ddot{\delta}^{jp^{f-1}}, \ddot{\delta}^{-jp^{f-1}}\}\langle\ddot{\delta}^{p^i+1}\rangle g$
$-$	$\ddot{\varphi}^i$	$2f/i$ odd	$\{\ddot{\delta}^j, \dots, \ddot{\delta}^{jp^{f-1}}\}\langle\ddot{\delta}^{p^i+1}\rangle g$

Remark 6.2.8 The main result of [21] is that $u(G) \geqslant 2$ for all almost simple linear groups $G \in \mathcal{A}_+$. Referring to Fig. 6.1, the automorphisms in columns 1, 2, 3, 6, 7 are considered in Sections 4, 5.1, 5.2, 3, 6 of [21], respectively. When q is odd and n is even, in Section 6, the authors consider both γ and $\delta_2\gamma$, but, in Section 5.2 where f/i is odd, only φ^i is considered, since it was claimed that $\ddot{\gamma}\ddot{\varphi}^i$ and $\ddot{\delta}_2\ddot{\gamma}\ddot{\varphi}^i$ were $\mathrm{Out}(T)$-conjugate, but we know that this does not hold by Lemma 6.2.4. The basis of this claim was [36, Theorem 7.2], which states that for a finite simple group of Lie type K and a field or graph-field automorphism $\phi \in \mathrm{Aut}(K)$, if $\phi' \in \phi\,\mathrm{Inndiag}(K)$ has the same order as ϕ, then ϕ and ϕ' are conjugate under $\mathrm{Inndiag}(K)$. While this

statement is true for prime order elements, it is false in general, as the example of $\gamma\varphi^i$ and $\delta_2\gamma\varphi^i$ in $\mathrm{Aut}(\mathrm{PSL}_n(q))$ when f/i is odd highlights. Since $\ddot{\delta}_2\ddot{\gamma}\ddot{\varphi}^i$ is not $\mathrm{Out}(T)$-conjugate to $\ddot{\gamma}\ddot{\varphi}^i$, the group $\langle T, \delta_2\gamma\varphi^i\rangle$ is not $\mathrm{Aut}(T)$-conjugate to $\langle T, \gamma\varphi^i\rangle$ and hence, by [7, Lemma 3], not isomorphic to $\langle T, \gamma\varphi^i\rangle$. Therefore, we must consider this case. In proving Theorem 6.3 in Sect. 6.6, we do exactly this.

As in Chap. 5, we can deal with some small cases by computation.

Proposition 6.2.9 *Let $G \in \mathcal{A}_-$ with socle $\mathrm{PSU}_n(q)$. Then $u(G) \geqslant 2$ if $n \in \{3, 4\}$ and $q \leqslant 9$, $n \in \{5, 6\}$ and $q \leqslant 4$ or $n \in \{7, 8\}$ and $q \leqslant 3$.*

6.3 Elements

In Sect. 5.3 we introduced types of elements of symplectic and orthogonal groups. We now define types of semisimple elements in unitary groups.

For this section, write $V = \mathbb{F}_{q^2}^n$ where $n \geqslant 3$ and $q = p^f$, and let $\alpha = \alpha_- \in \mathbb{F}_{q^2}^\times$ have order $q + 1$.

Definition 6.3.1 Let $n \geqslant 3$ be odd. An element $g \in \mathrm{GU}_n(q)$ has *type* $[n]_q^-$ if V is an irreducible $\mathbb{F}_{q^2}\langle g\rangle$-module and g has order $q^n + 1$ and determinant α.

Lemma 6.3.2 *Let $n \geqslant 3$ be odd. Then $\mathrm{GU}_n(q)$ has an element of type $[n]_q^-$.*

Proof Fix a field extension embedding $\pi\colon \mathrm{GU}_1(q^n) \to \mathrm{GU}_n(q)$. Note that

$$\mathrm{GU}_1(q^n) = \{(\mu) \in \mathrm{GL}_1(q^{2n}) \mid \mu^{q^n+1} = 1\} \cong C_{q^n+1}.$$

Let $N\colon \mathbb{F}_{q^{2n}}^\times \to \mathbb{F}_q^\times$ be the norm map. Let $H \leqslant \mathbb{F}_{q^{2n}}^\times$ have order $q^n + 1$. Since n is odd, $N(H)$ has order

$$(q^n + 1)/\left(\frac{q^{2n} - 1}{q^2 - 1}, q^n + 1\right) = q + 1,$$

so there exists a generator λ of H such that $N(\lambda) = \alpha$, and the element $g = \pi((\lambda))$ has order $q^n + 1$. Now the determinant of g is $N(\lambda) = \alpha$ and λ is an eigenvalue of g. Therefore, Lemma 5.3.1 implies that g is irreducible on V, so g has type $[n]_q^-$. □

Let $n = 2m$ be even. Then V admits a decomposition $\mathcal{D}(V)$

$$V = V_1 \oplus V_2 \quad \text{where} \quad V_1 = \langle e_1, \ldots, e_m\rangle \quad \text{and} \quad V_2 = \langle f_1, \ldots, f_m\rangle,$$

noting that V_1 and V_2 are totally singular m-spaces. The centraliser of $\mathcal{D}(V)$ is

$$(\mathrm{GU}_{2m}(q))_{(\mathcal{D}(V))} = \{(g_{ij}) \oplus (g_{ij}^q)^{-\mathsf{T}} \mid (g_{ij}) \in \mathrm{GL}_m(q^2)\}. \tag{6.10}$$

Definition 6.3.3 Let $n = 2m \geqslant 4$. An element $g \in \mathrm{GU}_n(q)$ has *type* $[n]_q^+$ if g has order $q^n - 1$, determinant α and centralises a decomposition $V = V_1 \oplus V_2$ where V_1 and V_2 are totally singular subspaces on which g^i acts irreducibly for all divisors i of $q + 1$.

Lemma 6.3.4 *Let $n \geqslant 4$ be even. Then $\mathrm{GU}_n(q)$ has an element of type $[n]_q^+$.*

Proof Fix a field extension embedding $\pi\colon\ \mathrm{GL}_1(q^{2m}) \to \mathrm{GL}_m(q^2)$ where we write $n = 2m$. Since $|\alpha| = q + 1$, we may write $\alpha = \mu^{1-q}$ for a generator μ of $\mathbb{F}_{q^2}^\times$. Let $N\colon \mathbb{F}_{q^{2m}}^\times \to \mathbb{F}_{q^2}^\times$ be the norm map and let $\lambda \in \mathbb{F}_{q^{2m}}^\times$ satisfy $N(\lambda) = \mu$.

Let $g = \pi((\lambda)) \oplus \pi((\lambda))^{-(q)\mathsf{T}}$, and note that g has order $q^n - 1$. The determinant of $\pi((\lambda))$ is $N(\lambda) = \mu$, so the determinant of g is $\mu\mu^{-q} = \alpha$.

Let i divide $q+1$. Now λ^i is an eigenvalue of $\pi(\lambda)^i$, and since $|\lambda^i| = (q^{2m}-1)/i$ is a primitive divisor of $q^{2m} - 1$, Lemma 5.3.1 implies that $\pi(\lambda)^i$ acts irreducibly on V_1, so g^i acts irreducibly on both V_1 and V_2. Therefore, g has type $[n]_q^+$. □

The following proof is based on the arguments in [13, Chapter 3].

Lemma 6.3.5 *Let $n = 2m \geqslant 4$. Let $g \in \mathrm{GU}_{2m}(q)$ have type $[n]_q^+$, centralising the decomposition $V = V_1 \oplus V_2$. For a divisor i of $q + 1$, the only $\mathbb{F}_{q^2}\langle g^i\rangle$-submodules of V are 0, V_1, V_2 and V.*

Proof Evidently, it suffices to prove the lemma when $i = q+1$, so write $h = g^{q+1}$. If m is even, then V_1 and V_2 are nonisomorphic, since h has different eigenvalues on these two submodules. Since V_1 and V_2 are irreducible $\mathbb{F}_{q^2}\langle h\rangle$-modules, the result follows from Lemma 2.3.3. Therefore, for the remainder of the proof we will assume that m is odd.

Consider V_1 and V_2 as copies of $\mathbb{F}_{q^m}$ where the action of h on V is given as $(v_1, v_2)h = (\mu v_1, \mu^{-q} v_2)$ for some $\mu \in \mathbb{F}_{q^{2m}}$ of order $(q^{2m} - 1)/(q + 1)$.

For a contradiction, let $0 < U < V$ be an $\mathbb{F}_{q^2}\langle h\rangle$-submodule different from V_1 and V_2. In particular, $U \cap V_1 = U \cap V_2 = 0$. Therefore, by Lemma 2.3.1, U is isomorphic to V_1 and V_2, so, in particular, U is m-dimensional. This means that for all $v_1 \in V_1$, there exists a unique $v_2 \in V_2$ such that $(v_1, v_2) \in U$. In this way, we can define a map $L\colon V_1 \to V_2$ as $L(v_1) = v_2$ where $(v_1, v_2) \in U$.

Fix $(1, u) \in U$. Since m is odd, $q + 1$ divides $q^m + 1$, so $\mathbb{F}_{q^m}^\times \leqslant \langle\mu\rangle$. Let $\langle\zeta\rangle = \mathbb{F}_{q^m}^\times$, noting that $\zeta \neq -1$ since $m \geqslant 2$. Now $(\zeta, \zeta^{-q}u) \in U$, so $(1 + \zeta, (1 + \zeta^{-q})u) \in U$. Now $1 + \zeta \in \mathbb{F}_{q^m} \leqslant \langle\mu\rangle$, so $(1 + \zeta, (1 + \zeta)^{-q}u) \in U$. Therefore, $1 + \zeta^{-q} = (1 + \zeta)^{-q} = (1 + \zeta^q)^{-1}$. This implies that $1 + \zeta + \zeta^2 = 1$. Therefore $\zeta^3 = 1$. However, $|\zeta| = q^m - 1$, so $q^m - 1 \leqslant 3$, which implies that $q = m = 2$. We can check that $g \in \mathrm{SU}_4(2)$ of order $(4^2 - 1)/(2 + 1) = 5$ does not stabilise such a subspace U, which gives a contradiction. □

6.4 Case I: Semilinear Automorphisms

In this section, we prove Theorems 6.1 and 6.2 in Case I. To this end, write $G = \langle T, \theta \rangle \in \mathcal{A}_-$ where $T = \mathrm{PSU}_n(q)$ and $\theta \in \mathrm{Aut}(T) \setminus \mathrm{Inndiag}(T)$.

We separate into two cases, which will be considered in Sects. 6.4.1 and 6.4.2:

(a) $G \cap \langle \mathrm{PGU}_n(q), \gamma \rangle \leqslant \mathrm{PGU}_n(q)$
(b) $G \cap \langle \mathrm{PGU}_n(q), \gamma \rangle \nleqslant \mathrm{PGU}_n(q)$.

6.4.1 Case I(a)

As in Case I of Chap. 5, Shintani descent (see Chap. 3) is the central tool in the identification of the element $t\theta$. Let us fix our notation for Case I(a).

Notation 6.4.1 Write $q = p^f$ where $f \geqslant 2$. Let $V = \mathbb{F}_{q^2}^n$.
Fix the basis $\mathcal{B}$ in (2.10).
Fix the simple algebraic group $X = \mathrm{PSL}_n(\overline{\mathbb{F}}_p)$.
Fix the Frobenius endomorphism $\varphi = \varphi_{\mathcal{B}}$, the standard graph automorphism $\gamma = \gamma_{\mathcal{B}}$ and the antidiagonal element $\delta = \delta_-$ (see Definitions 2.5.1 and 6.2.1).
Fix $\alpha = \alpha_- \in \mathbb{F}_{q^2}^\times$ satisfying $|\alpha| = q + 1$.

Our approach is like that for minus-type orthogonal groups in Sect. 5.4.1. By Proposition 6.2.6, we can assume that $\theta \in \mathrm{PGU}_n(q)\varphi^i$ where $2f/i$ is odd. Therefore, i is even and for $j = i/2$ we have $2f/(2f, f + j) = 2f/(2f, i)$, so we will work with $\theta = \theta_0\gamma\varphi^j$ for some $\theta_0 \in \mathrm{PGU}_n(q)$, noting that j divides f and $f/j = 2f/i$ is odd.

Notation 6.4.1. (continued) Write $q = q_0^e$ where $e = f/j = 2f/i$.
Let $\sigma = \gamma\varphi^j$ and let F be the Shintani map of (X, σ, e), so

$$F \colon \{(g\sigma)^{X_{\sigma^e}} \mid g \in X_{\sigma^e}\} \to \{x^{X_\sigma} \mid x \in X_\sigma\}.$$

Note $X_{\sigma^e} = \mathrm{PGU}_n(q) = \mathrm{Inndiag}(T)$ and $X_\sigma = \mathrm{PGU}_n(q_0) = \mathrm{Inndiag}(T_0)$, where $T_0 = \mathrm{PSU}_n(q_0)$.
Fix the antidiagonal element $\delta_0 \in \mathrm{PGU}_n(q_0)$ corresponding to $\delta \in \mathrm{PGU}_n(q)$.
Let $\alpha_0 \in \mathbb{F}_{q_0^2}^\times$ satisfy $|\alpha_0| = q_0 + 1$

Let $N\colon \mathbb{F}_{q^2} \to \mathbb{F}_{q_0^2}$ be the norm map. Then

$$|N(\alpha)| = (q+1)/\left(\tfrac{q^2-1}{q_0^2-1}, q+1\right) = q_0 + 1$$

noting that e is odd. Since the normal closure of $\langle \mathrm{PSU}_n(q), \varphi^i \rangle$ in $\langle \mathrm{PGU}_n(q), \varphi^i \rangle$ is $\langle \mathrm{PSU}_n(q), \delta^{q_0+1}, \varphi^i \rangle$, there is a well-defined bijection between the normal unions of cosets of $\mathrm{PSU}_n(q)$ in $\mathrm{PGU}_n(q)\sigma$ and the (necessarily normal) cosets of $\mathrm{PSU}_n(q_0)$ in $\mathrm{PGU}_n(q_0)$. The following demonstrates that the Shintani map preserves this bijection (compare with [21, Lemmas 4.2 and 5.3]).

Lemma 6.4.2 *Let $x \in X_\sigma$. Write $\det(x) = \lambda^{1+q_0^2+\cdots+q_0^{2e-2}}$, where $\lambda \in \mathbb{F}_{q^2}^\times$ with $\lambda^{q+1} = 1$. There exists $g \in X_{\sigma^e}$ such that $F((g\sigma)^{X_{\sigma^e}}) = x^{X_\sigma}$ and $\det(g) = \lambda$.*

Proof There exists $h \in X_{\sigma^e}$ such that $x^{X_\sigma} = F((h\sigma)^{X_{\sigma^e}}) = (a^{-1}(h\sigma)^e a)^{X_\sigma}$. Now

$$\lambda^{1+q_0^2+\cdots+q_0^{2(e-1)}} = \det(x) = \det(a^{-1}(h\sigma)^e a) = \det(h)^{1+q_0^2+\cdots+q_0^{2(e-1)}}.$$

Therefore, $\det(h) = \lambda \cdot \mu^{q_0^2-1}$ for some $\mu \in \mathbb{F}_{q^2}^\times$ such that $\mu^{q+1} = 1$. Now let $z \in \mathrm{PGU}_n(q) = X_{\sigma^e}$ satisfy $\det(z) = \mu^{-q_0^2}$ and write $g = zhz^{-\sigma^{-1}}$. Then

$$g\sigma = (zhz^{-\sigma^{-1}})\sigma = (h\sigma)^{z^{-1}} \in (h\sigma)^{X_{\sigma^e}},$$

so $F((g\sigma)^{X_{\sigma^e}}) = F((h\sigma)^{X_{\sigma^e}}) = x^{X_\sigma}$, and $g \in X_{\sigma^e}$ satisfies the statement since

$$\det(g) = \det(z)\det(h)\det(z)^{-q_0^{2e-2}} = \mu^{-q_0^2}\lambda\mu^{q_0^2-1}\mu = \lambda.$$

This completes the proof. □

Proposition 6.4.3 *Let $T \in \mathcal{T}_-$ and let $\theta = \delta^\ell \gamma \varphi^j$ where $1 \leqslant \ell \leqslant (q+1, n)$ and j is a proper divisor of f. Write $y = y_0^\ell$, for the element $y_0 \in \mathrm{PGU}_n(q_0)$ in Fig. 6.2. Then there exists $t \in T$ such that $(t\theta)^e$ is X-conjugate to y.*

Proof Note that $y \in \mathrm{PGU}_n(q_0)$ and $\det(y) = \alpha_0^\ell$. Without loss of generality, assume that $\alpha_0 = \alpha^{1+q_0^2+\cdots+q_0^{2e-2}}$, so by Lemma 6.4.2, there exists $g \in \mathrm{PGU}_n(q)$ such that

Fig. 6.2 Case I(a): The element y_0 for the automorphism θ

n	y_0
$n \geqslant 7$ odd	$J_2 \perp [n-2]^-_{q_0}$
$n \geqslant 6$ even	$J_2 \perp [n-2]^+_{q_0}$
5	$[5]^-_{q_0}$
4	$I_1 \perp [3]^-_{q_0}$
3	$[3]^-_{q_0}$

Fig. 6.3 Case I(a): Description of $\mathcal{M}(G, t\theta)$

	type of H	$m(H)$	conditions
$\mathcal{C}_1$	$\mathrm{GU}_2(q) \times \mathrm{GU}_{n-2}(q)$	1	
	P_1	1	
	P_{m-1}	2	$n = 2m$
	P_m	2	$n = 2m$
$\mathcal{C}_2$	$\mathrm{GU}_{n/k}(q) \wr S_k$	N	$k > 1,\ k \mid n$
	$\mathrm{GL}_m(q^2)$	N	$n = 2m$
$\mathcal{C}_5$	$\mathrm{GU}_n(q^{1/k})$	N	k odd prime, $k \mid f$
	$\mathrm{Sp}_n(q)$	N	n even

Note: $N = |C_{\mathrm{PGU}_n(q_0)}(y)|$

$\det(g) = \alpha^\ell$ and $F(g\sigma) = y$. Therefore, we may write $g = t\delta^\ell$ where $t \in T$, so $g\sigma = t\delta^\ell\varphi^i = t\theta$. Now $y = a^{-1}(t\theta)^e a = F(t\theta)$ for some $a \in X$, as claimed. □

Proposition 6.4.4 *Assume that $n \geqslant 6$. The maximal subgroups of G which contain $t\theta$ are listed in Fig. 6.3, where $m(H)$ is an upper bound on the multiplicity of the subgroups of type H in $\mathcal{M}(G, t\theta)$.*

Proof Let $H \in \mathcal{M}(G, t\theta)$ and note that $T \not\leqslant H$. First assume that $H \notin \mathcal{C}_1$. A power of $t\theta$ is X-conjugate to y, a power of which is $J_2 \perp I_{n-2}$. Therefore, Proposition 4.2.5 implies that $H \in \mathcal{C}_2 \cup \mathcal{C}_5$, noting that $q > p$. Moreover, if q is odd, then H does not have type $\mathrm{O}_n^\varepsilon(q)$ as orthogonal groups do not contain elements of Jordan form $[J_2, J_1^{n-2}]$. All other possible types of such subgroups are given in Fig. 6.3. All geometric subgroups of G of a given type are $\langle X_{\sigma^e}, \sigma\rangle$-conjugate by [50, Theorem 4.0.2] and the upper bound on the multiplicity $m(H)$ is provided by Proposition 3.3.4.

Now assume that $H \in \mathcal{C}_1$. By Lemma 2.3.3 (in conjunction with Lemma 6.3.5), the reducible subgroups of X_σ that contain y are one of type $\mathrm{GU}_2(q_0) \perp \mathrm{GU}_{n-2}(q)$, one of type P_1 and if n is even also two of type $P_{n/2-1}$ and two of type $P_{n/2}$. By applying Lemma 3.3.2 with Y as the (connected) subgroup of X of type $\mathrm{GL}_k(\overline{\mathbb{F}}_p) \times \mathrm{GL}_{n-k}(\overline{\mathbb{F}}_p)$ or $P_{k,n-k}$ for each $1 \leqslant k < n/2$ (in the notation of [50, Section 4.1]) we conclude that the reducible subgroups of G that contain $t\theta$ are those in the statement. □

Proposition 6.4.5 *Let $G = \langle T, \theta\rangle \in \mathcal{A}$ where $T = \mathrm{PSU}_n(q)$. In Case I(a), $u(G) \geqslant 2$ and as $q \to \infty$ we have $u(G) \to \infty$.*

Proof Let $x \in G$ have prime order. As usual, write $\mathcal{M}(G, t\theta)$ for the set of maximal subgroups of G that contain $t\theta$.

First assume that $n \geqslant 6$. Then $\mathcal{M}(G, t\theta)$ is described by Proposition 6.4.4. By Lemma 2.3.4, $|C_{X_\sigma}(y)| \leqslant q_0^{n-1} + q_0$. Write $d(k)$ for the number of divisors of k.

Upper bounds on the fixed point ratios for subspace actions are given by Theorem 4.1.1. Now assume that $H \leqslant G$ is a maximal irreducible subgroup. If $n \geqslant 7$, then Proposition 4.2.3 implies that $\mathrm{fpr}(x, G/H) < 2q^{-(n-3+2/n)}$ and Theorem 4.2.4 implies that the same conclusion holds for $n = 6$ too.

Applying Lemma 3.3.4, if $n \geqslant 7$ is odd, then

$$P(x, t\theta) < \frac{1}{q^2} + \frac{1}{q^4} + \frac{1}{q^{n/2-3/2}} + \frac{1}{q^{n-3}} + \frac{4}{q^{n-2}} + \frac{1}{q^n} + (d(n) + \log\log q + 1) \cdot (q_0^{n-1} + q_0) \cdot \frac{2}{q^{n-3+2/n}} < \frac{1}{2}$$

and $P(x, t\theta) \to 0$ and $q \to \infty$ (recall that $q = q_0^e$ where $e = f/j \geqslant 3$ is odd). Similarly, if $n \geqslant 6$ is even, then

$$P(x, t\theta) < \frac{1}{q^2} + \frac{1}{q^4} + \frac{2}{q^{n-4}} + \frac{5}{q^{n/2-1}} + \frac{11}{q^{n-2}} + \frac{5}{q^{n-1}} + (d(n) + \log\log q + 1) \cdot (q_0^{n-1} + q_0) \cdot \frac{2}{q^{n-3+2/n}} < \frac{1}{2}$$

and $P(x, t\theta) \to 0$ as $q \to \infty$.

Next assume that $n \in \{3, 5\}$. We begin by determining the possible types of subgroups in $\mathcal{M}(G, t\theta)$. First consider reducible subgroups. Since y is not contained in any reducible subgroups of $\mathrm{PGU}_n(q_0)$, by Lemma 3.3.2, $t\theta$ is not contained in any reducible subgroups of G. Therefore, if $H \in \mathcal{M}(G, t\theta)$, then consulting [8, Tables 8.5, 8.6, 8.20 and 8.21], we see that H has one of the following types: $\mathrm{GU}_1(q) \wr S_n$, $\mathrm{GU}_1(q^n)$ and $\mathrm{GU}_n(q^{1/k})$ for k dividing f.

With the bound on fixed point ratios from Theorem 4.2.4, if $n = 5$, then

$$P(x, t\theta) \leqslant (2 + \log\log q) \cdot \frac{q_0^5 + 1}{q_0 + 1} \cdot \frac{4}{3q^4} < \frac{1}{2},$$

and if $n = 3$, then by Proposition 6.2.9 we assume that $q \neq 8$ and we obtain

$$P(x, t\theta) \leqslant (2 + \log\log q) \cdot \frac{q_0^3 + 1}{q_0 + 1} \cdot \frac{1}{q^2 - q + 1} < \frac{1}{2}.$$

In both cases, $P(x, t\theta) \to 0$ as $q \to \infty$.

Finally assume that $n = 4$. By Proposition 6.2.9, we may assume that $q \neq 8$. Since y is contained in a unique reducible subgroup of $\mathrm{PGU}_4(q_0)$ (of type $\mathrm{GU}_3(q_0)$), by Lemma 3.3.2, we know that $t\theta$ is contained in a unique reducible subgroup of G (of type $\mathrm{GU}_3(q)$). From [8, Table 8.10 and 8.11], the types of irreducible maximal subgroups of G are $\mathrm{GU}_1(q) \wr S_4$, $\mathrm{GU}_2(q) \wr S_2$, $\mathrm{GL}_2(q^2)$, $\mathrm{Sp}_4(q)$, $\mathrm{O}_4^{\pm}(q)$ and $\mathrm{GU}_4(q^{1/k})$ for k dividing f. Notice that y is not contained in a subgroup of type $\mathrm{Sp}_4(q)$ or $\mathrm{O}_4^{\pm}(q)$ since such groups do not contain semisimple elements with a one-dimensional 1-eigenspace. Therefore, using the fixed point ratio bounds in (4.1) (for the subspace subgroup) and Theorem 4.2.4 (for the nonsubspace subgroups), we obtain

$$P(x, t\theta) < \frac{4}{3q} + (3 + \log\log q) \cdot (q_0^3 + 1) \cdot \frac{1}{q^2 - q + 1} < \frac{1}{2}$$

and $P(x, t\theta) \to 0$ as $q \to \infty$. The desired result now follows by Lemma 2.1.1. □

6.4.2 *Case I(b)*

As in Sect. 5.4.2, we need a variant on Shintani descent to identify an element of $T\theta$. Let us fix our notation for this section.

Notation 6.4.6 Write $q = p^f$ where $f \geqslant 1$. Let $V = \mathbb{F}_{q^2}^n$.
Fix the basis $\mathcal{B}$ from (2.10).
Fix the simple algebraic group $X = \mathrm{PSL}_n(\overline{\mathbb{F}}_p)$.
Fix the Frobenius endomorphism $\varphi = \varphi_{\mathcal{B}}$ and the standard graph automorphism $\gamma = \gamma_{\mathcal{B}}$ (see Definition 2.5.1).
If $(n, q+1)$ is even, then fix the antidiagonal element $\delta_2 = \delta^{\frac{q+1}{(q+1)_2}}$, where δ is given in Definition 6.2.1, so $|\delta_2| = (n, q+1)_2$ (see Remark 6.2.5).

By Proposition 6.2.6, we can assume that $\theta = \varphi^i$ or, if $(n, q+1)$ is even, $\theta = \delta_2\varphi^i$ where i is a proper divisor of f.

Notation 6.4.6. (continued) Write $q = q_0^e$ where $e = f/i$.
Fix the Steinberg endomorphism σ and the graph automorphism ρ according to the following two cases. In both cases, write $Z = C_X(\rho)^\circ$.

case	θ	e	σ	ρ
(i)	φ^i	all	φ^i	γ
	$\delta_2\varphi^i$	even	$\delta_2\varphi^i$	γ
(ii)	$\delta_2\varphi^i$	odd	$\delta_2\varphi^i$	$\delta_2\gamma$

Proposition 6.4.7 *The automorphism ρ is an involution that commutes with σ and the isomorphism type of Z_σ is given in Fig. 6.4.*

Fig. 6.4 Case I(b): The group Z_σ

n	q	Z_σ	condition
odd	even	$\mathrm{PSp}_{n-1}(q_0)$	
	odd	$\mathrm{PSO}_n(q_0)$	
even	even	$\mathrm{PSp}_n(q_0)$	
	odd	$\mathrm{PGSp}_n(q_0)$	case (i)
		$\mathrm{PDO}_n^\eta(q_0)$	case (ii)

Note: in the final row, $\eta = (-)^{\frac{n(q+1)}{4}+1}$

Proof If $\theta = \varphi^i$, then $\sigma = \varphi^i$ and $\rho = \gamma$, so clearly $|\rho| = 2$ and $\rho^\sigma = \rho$, and Z_σ is given by (6.5). For the remainder of the proof, we will assume that $\theta = \delta_2\varphi^i$, so, in particular, $n = 2m$ is even and q is odd.

First assume that e is even, so $\sigma = \delta_2\varphi^i$ and $\rho = \gamma$. Clearly ρ is an involution. Since e is even, $q \equiv 1 \pmod 4$, so $|\delta_2| = 2$. Therefore, $\delta_2^\varphi = \delta_2^\gamma = \delta_2$, which implies that σ and ρ commute. By (6.4), $Z = \mathrm{PGSp}_n(\overline{\mathbb{F}}_p)$. Since Z is connected and $\delta_2 \in Z$, we know that $Z_{\delta_2\varphi^i} \cong Z_{\varphi^i} = \mathrm{PGSp}_n(q_0)$.

Now assume that e is odd, so $\sigma = \delta_2\varphi^i$ and $\rho = \delta_2\gamma$. Since e is odd, the quotient $(q+1)/(q_0+1) = q_0^{e-1} - q_0^{e-2} + \cdots - q_0 + 1$ is odd, so $|\delta_2| = (n, q+1)_2 = (n, q_0+1)_2$. In particular, $\delta_2^{\varphi^i} = \delta_2^\gamma = \delta_2^{-1}$, so $(\delta_2\gamma)^2 = 1$ and $(\delta_2\varphi^i)(\delta_2\gamma) = \varphi^i\gamma = (\delta_2\gamma)(\delta_2\varphi^i)$. Finally, by Remark 6.2.5(iv)

$$\begin{aligned} C_X(\rho)_\sigma &= \{x \in X \mid x^{\delta_2\varphi^i} = x \text{ and } x^{\delta_2\gamma} = x\} \\ &= \{x \in X \mid x^{\gamma\varphi^i} = x \text{ and } x^{\delta_2\gamma} = x\} \\ &= C_{\mathrm{PGU}_n(q_0)}(\delta_2\gamma) \\ &= \mathrm{PGO}_n^\eta(q), \end{aligned}$$

where $\eta = (-)^{\frac{n(q_0+1)}{2}+1} = (-)^{\frac{n(q+1)}{2}+1}$ as e is odd. Therefore, $Z_\sigma = \mathrm{PDO}_n^\eta(q_0)$. □

Proposition 6.4.8 *Let y be the element from Fig. 6.5. Then there exists $t \in T$ that commutes with ρ such that $(t\theta)^e$ is X-conjugate to $y\rho$.*

Proof By Proposition 6.4.7, ρ is an involution that commutes with σ, so $(\rho\sigma)^{2e} = \sigma^{2e}$ and $Z = C_X(\rho)^\circ$ is σ-stable. Moreover, from the structure of Z_σ given in Fig. 6.4, it is easy to see that Z_σ contains elements of the type given in Fig. 6.5. Therefore, Lemma 3.4.1(i) implies that there exists $t \in Z_{\sigma^e} \leqslant X_{\gamma\sigma^e}$ such that $(t\sigma)^e$ is X-conjugate (indeed Z-conjugate) to $y\rho$. If n is odd or q is even, then $Z_{\sigma^e} \leqslant T$, so we, in fact, have $t \in T$. Now assume that n is even and q is odd. In case (ii), by applying Lemma 5.4.3 to the Shintani map of (Z, σ, e), we see that $t \in \mathrm{PSO}_n^\pm(q) \leqslant \mathrm{PSU}_n(q)$ since $y \in \mathrm{PSO}_n^\eta(q_0)$. Similarly, in case (i), we apply [43, Proposition 2.4] (see Example 3.2.4) to conclude that $t \in \mathrm{PSp}_n(q) \leqslant \mathrm{PSU}_n(q)$ since $y \in \mathrm{PSp}_n(q_0)$. □

Fig. 6.5 Cases I(b) and II(b): The element y

Generic case: $n \geqslant 7$ odd or $n \geqslant 12$ even			
case	n	q	y
(i)	odd	even	$(n-3)^-_{q_0} \perp (2)^-_{q_0} \perp I_1$
		odd	$(n-3)^-_{q_0} \perp J_3$
	even		$(n-2)^-_{q_0} \perp (2)^-_{q_0}$
(ii)	0 (mod 4)		$(n-2)^+_{q_0} \perp (2)^-_{q_0}$
	2 (mod 4)	1 (mod 4)	$(n-2)^-_{q_0} \perp (2)^-_{q_0}$
		3 (mod 4)	$(\frac{n}{2}+1)^-_{q_0} \perp (\frac{n}{2}-3)^-_{q_0} \perp (2)^-_{q_0}$

Specific cases: $n \leqslant 5$ odd or $n \leqslant 10$ even		
case	n	y
(i)	5	$(4)^-_{q_0} \perp I_1$
	3	$g \perp I_1$
	even	$(n)^-_{q_0}$
(ii)		$(n)^{\eta}_{q_0}$

Note: when $n = 3$, let $g \in \mathrm{SO}_2^-(q_0)$ have order $q_0 + 1$

Remark 6.4.9 Let $g \in \mathrm{SO}_2^-(q_0)$ have type $(2)^-_{q_0}$. If q_0 is not Mersenne, then g and g^2 have odd order and two distinct eigenvalues. Now assume that q_0 is Mersenne. In this case, $|g| = q_0 + 1$, which is a power of two. Therefore, g has two distinct eigenvalues, and g^2 also has two distinct eigenvalues, unless $q_0 = 3$, in which case $g^2 = -I_2$. For this reason, in several arguments in this section (and those that follow), the case where $q_0 = 3$ will require special attention.

Proposition 6.4.10 *Assume that $n \geqslant 7$ is odd or $n \geqslant 12$ is even. Then the maximal subgroups of G that contain $t\theta$ are listed in Fig. 6.6, where $m(H)$ is an upper bound on the multiplicity of the subgroups of type H in $\mathcal{M}(G, t\theta)$.*

Proof First assume that $H \notin \mathcal{C}_1$. An X-conjugate of $(t\theta)^{2e}$ is $(y\gamma)^2 = y^2$. If n is odd or q_0 is not Mersenne, then a suitable power of y^2 has type $J_3 \perp I_{n-3}$ or $(2)^-_{q_0} \perp I_{n-2}$, so Proposition 4.2.5, implies that H appears in Fig. 6.6. Now assume that n is even and q_0 is Mersenne. Here the order of $(2)^-_{q_0}$ is $q_0 + 1$, which is a power of two, so a suitable power of y^2 is $[\lambda, \lambda^{q_0}] \perp I_{n-2}$ where $|\lambda| = (q_0 + 1)/2$. Again Proposition 4.2.5 gives the possible types for H but note that when $q_0 = 3$ we have $\lambda = \lambda^{q_0} = -1$, so the type $\mathrm{GU}_2(q) \otimes \mathrm{GU}_{n/2}(q)$ also arises. In both cases, all maximal subgroups of a given type are $\langle X_{\gamma\sigma^e}, \sigma\rangle$-conjugate, so Lemma 3.4.1(ii)(a) gives $m(H)$, noting that Proposition 6.4.7 implies $(\rho\sigma^e)^2 = \sigma^{2e}$.

We now assume that $H \in \mathcal{C}_1$. We will provide the details when n and q are even; the other cases are similar. By Lemma 2.3.3, the proper nonzero subspaces of $\mathbb{F}_{q_0}^n$ stabilised by y are U_0 and $U_0^\perp$, which are orthogonal nondegenerate subspaces of dimensions 2 and $n-2$. Moreover, these are also the proper nonzero subspaces stabilised by y^2. We now apply Lemma 3.4.1(ii)(b). Since the unique γ-stable $\mathcal{C}_1$ subgroup of $\mathrm{GL}_n(q_0)$ containing y^2 has type $\mathrm{GL}_2(q_0) \times \mathrm{GL}_{n-2}(q_0)$, there is a unique

	type of H	$m(H)$	conditions
$\mathcal{C}_1$	P_1	1	n odd, q odd
	P_2	1	n odd, q odd
	$P_{\frac{n}{2}-1}$	2	case (ii), $n \equiv 0 \pmod 4$
	$\mathrm{GU}_1(q) \times \mathrm{GU}_{n-1}(q)$	1	n odd, q even
		4	n even, $q_0 = 3$
	$\mathrm{GU}_2(q) \times \mathrm{GU}_{n-2}(q)$	1	n even or q even
	$\mathrm{GU}_3(q) \times \mathrm{GU}_{n-3}(q)$	1	n odd
	$\mathrm{GU}_{\frac{n}{2}-3}(q) \times \mathrm{GU}_{\frac{n}{2}+3}(q)$	1	case (ii), $n \equiv 2 \pmod 4$, $q \equiv 3 \pmod 4$
	$\mathrm{GU}_{\frac{n}{2}-2}(q) \times \mathrm{GU}_{\frac{n}{2}+2}(q)$	4	case (ii), $n \equiv 2 \pmod 4$, $q_0 = 3$
	$\mathrm{GU}_{\frac{n}{2}-1}(q) \times \mathrm{GU}_{\frac{n}{2}+1}(q)$	1	case (ii), $n \equiv 2 \pmod 4$, $q \equiv 3 \pmod 4$
$\mathcal{C}_2$	$\mathrm{GU}_{\frac{n}{k}}(q) \wr S_k$	N	$k > 1$, $k \mid n$
	$\mathrm{GL}_{\frac{n}{2}}(q^2)$	N	n even
$\mathcal{C}_4$	$\mathrm{GU}_2(q) \otimes \mathrm{GU}_{\frac{n}{2}}(q)$	N	n even, $q_0 = 3$
$\mathcal{C}_5$	$\mathrm{GU}_n(q^{1/k})$	N	k odd prime, $k \mid f$
	$\mathrm{Sp}_n(q)$	N	n even
	$\mathrm{O}_n^{\upsilon}(q)$	N	q odd, $\upsilon \in \{+, \circ, -\}$

Note: $N = |C_{\mathrm{PGL}_n(q_0)}(y^2)|$

Fig. 6.6 Case I(b): Description of $\mathcal{M}(G, t\theta)$

γ-stable reducible subgroup of $\langle \mathrm{PGL}_n(q^2), \sigma \rangle$ containing $t\theta$ and moreover it has type $\mathrm{GL}_2(q^2) \times \mathrm{GL}_{n-2}(q^2)$. Since any reducible subgroup of $\widetilde{G} = \langle \mathrm{PGU}_n(q), \sigma \rangle$ containing $t\theta$ gives rise to an γ-stable subgroup of $\langle \mathrm{PGL}_n(q^2), \sigma \rangle$ containing $t\theta$, we deduce that the unique reducible subgroup of $\widetilde{G}$ containing $t\theta$ has type $\mathrm{GU}_2(q) \times \mathrm{GU}_{n-2}(q)$, so the same conclusion holds for G, as we claim in Fig. 6.6. □

We obtain a more precise bound in one case.

Proposition 6.4.11 *Assume that $T = \mathrm{PSU}_4(q)$ and e is even. Then $t\theta$ is contained in at most $k(q_0 + 1)$ maximal subgroups of G of type $\mathrm{Sp}_4(q)$ where $k = 2$ if $q_0 \equiv 3 \pmod 4$ and $k = 1$ otherwise.*

Proof Let H be a maximal subgroup of G of type $\mathrm{Sp}_4(q)$ and let m be the number of maximal subgroups of G of type $\mathrm{Sp}_4(q)$ that contain $t\theta$. Recall the algebraic groups $X = \mathrm{PGL}_4(\overline{\mathbb{F}}_p)$ and $Z = \mathrm{PGSp}_4(\overline{\mathbb{F}}_p)$, the Frobenius endomorphisms $\sigma \in \{\varphi^i, \delta_2\varphi^i\}$ and the graph automorphism $\rho = \gamma$. We have $G = \langle \mathrm{PSU}_4(q), \theta \rangle \leqslant \langle X_{\gamma\sigma^e}, \widetilde{\sigma} \rangle = \widetilde{G}$ and we may assume that $H \leqslant \langle \mathrm{PGSp}_4(q), \theta \rangle = \langle Z_{\sigma^e}, \widetilde{\sigma} \rangle = \widetilde{H}$. All subgroups of G of type $\mathrm{Sp}_4(q)$ are $\widetilde{G}$-conjugate, so, by Lemma 2.1.2, $(t\theta)^{\widetilde{G}} \cap H$ is the disjoint union $\bigcup_{i=1}^{k} (t_i\theta)^{\widetilde{H}}$ and

$$m \leqslant \frac{|\widetilde{G}|}{|\widetilde{H}|} \frac{|(t\theta)^{\widetilde{G}} \cap H|}{|(t\theta)^{\widetilde{G}}|} = \sum_{i=1}^{k} \frac{|C_{\widetilde{G}}(t\theta)|}{|C_{\widetilde{H}}(t_i\theta)|}.$$

Let $s\theta \in (t\theta)^{\widetilde{G}} \cap H$. Let E and F be the Shintani maps of $(X, \sigma, 2e)$ and (Z, σ, e) respectively. Since $s\theta$ is $\langle \mathrm{PGL}_4(q^2), \theta\rangle$-conjugate to $t\theta$, by Shintani descent, $E(s\theta) = F(s\theta)^2$ is $\mathrm{PGL}_4(q_0)$-conjugate to $E(t\theta) = F(t\theta)^2 = y^2$ (see Lemma 3.2.5). Now y^2 is semisimple, so by [20, Lemma 3.4.2], $F(s\theta)^2$ is $\mathrm{PGSp}_4(q_0)$-conjugate to y^2. The eigenvalues of y are $\lambda_1 = \lambda$, $\lambda_2 = \lambda^{q_0}$, $\lambda_3 = \lambda^{q_0^2}$, $\lambda_4 = \lambda^{q_0^3}$ where $\lambda \in \mathbb{F}_{q_0^4}^{\times}$ satisfies $|\lambda| = \Phi_4^*(q_0)$. Therefore, the eigenvalues of y^2, and hence $F(s\theta)^2$ are $\lambda_1^2, \lambda_2^2, \lambda_3^2, \lambda_4^2$.

For now assume that $q_0 \not\equiv 3 \pmod 4$. In this case, modulo scalars, the eigenvalues of $F(s\theta)$ are $\lambda_1, \lambda_2, \lambda_3, \lambda_4$, so $F(s\theta)$ is $\mathrm{PGSp}_4(q_0)$-conjugate to y. Therefore, by Shintani descent, $s\theta$ is $\widetilde{H}$-conjugate to $t\theta$. In this case, write $y_1 = y$ and $t_1 = t$.

Now assume that $q_0 \equiv 3 \pmod 4$. Here $F(s\theta) \in y_1^{\mathrm{PGSp}_4(q_0)} \cup y_2^{\mathrm{PGSp}_4(q_0)}$ where $y_1 = y = [\lambda_1, \lambda_2, \lambda_3, \lambda_4]$ and $y_2 = [\xi\lambda_1, -\xi\lambda_2, \xi\lambda_3, -\xi\lambda_4]$ where $\xi \in \mathbb{F}_{q_0^2}^{\times}$ satisfies $|\xi| = 4$, and we conclude that $s\theta$ is $\widetilde{H}$-conjugate to $t_1\theta$ or $t_2\theta$, where $F(t_1\theta) = y_1$ and $F(t_2\theta) = y_2$.

By Theorem 3.1.4 (in conjunction with Remark 3.2.6)

$$|C_{\widetilde{G}}(t\theta)| \leqslant |C_{\langle \mathrm{PGL}_4(q^2), \theta\rangle}(t\theta)| = 2e|C_{\mathrm{PGL}_4(q_0)}(y)| = 2e(q_0+1)(q_0^2+1)$$

and for $i \in \{1, 2\}$ we have $|C_{\widetilde{H}}(t_i\theta)| = 2e|C_{\mathrm{PGSp}_4(q_0)}(y_i)| = e(q_0^2+1)$.

Therefore, letting $k = 2$ if $q_0 \equiv 3 \pmod 4$ and $k = 1$ otherwise, we obtain

$$m \leqslant \sum_{i=1}^{k} \frac{|C_{\widetilde{G}}(t\theta)|}{|C_{\widetilde{H}}(t_i\theta)|} = k(q_0+1).$$

This completes the proof. □

We now establish the main result of this section.

Proposition 6.4.12 *Let $G = \langle T, \theta\rangle \in \mathcal{A}$ where $T = \mathrm{PSU}_n(q)$. In Case I(b), $u(G) \geqslant 2$ and as $q \to \infty$ we have $u(G) \to \infty$.*

Proof Let $x \in G$ have prime order. As usual, we apply Lemma 2.1.1. Bounds on the relevant fixed point ratios are given in Theorem 4.1.1 (for subspace actions) and Proposition 4.2.3 and Theorem 4.2.4 (for nonsubspace actions). As usual, we write $d(k)$ for the number of divisors of k.

First assume that $n \geqslant 7$ is odd and q is even. In this case,

$$|C_{\mathrm{PGL}_n(q_0)}(y^2)| = (q_0^2-1)(q_0^{n-3}-1),$$

so by Proposition 6.4.10,

$$P(x,t\theta) < \frac{1}{q^2}+\frac{1}{q^4}+\frac{1}{q^6}+\frac{8}{q^{n-3}}+\frac{1}{q^{n-1}}+\frac{3}{q^n}$$
$$+ (d(n)-1+\log\log q)\cdot(q_0^{n-3}-1)(q_0^2-1)\cdot\frac{2}{q^{n-3+2/n}} < \frac{1}{2}$$

unless $n=7$ and $q=4$. In this remaining case, there is one class of nonsubspace subgroups (of type $\mathrm{GU}_1(q)\wr S_7$), and by replacing $(d(n)-1+\log\log q)$ with 1 we obtain the desired result.

Next assume $n \geqslant 7$ and q are odd. Here $|C_{\mathrm{PGL}_n(q_0)}(y^2)| = q_0^2(q_0^{n-3}-1)$ and, by Proposition 6.4.10,

$$P(x,t\theta) < \frac{1}{q^2}+\frac{1}{q^4}+\frac{1}{q^6}+\frac{2}{q^{n/2}}+\frac{3}{q^{n-3}}+\frac{4}{q^{n-1}}+\frac{1}{q^n}$$
$$+ (d(n)-1+\log\log q)\cdot(q_0^{n-3}-1)q_0^2\cdot\frac{2}{q^{n-3+2/n}} < \frac{1}{2}$$

unless $e=2$ and $n=7$, when there are two classes of nonsubspace subgroups in $\mathcal{M}(G,t\theta)$ (of types $\mathrm{GU}_1(q)\wr S_7$ and $\mathrm{O}_n(q)$) and replacing $(d(n)+\log\log q)$ with 2 gives the result.

Now assume that $n \in \{3,5\}$. By Proposition 6.2.9, we will assume that $(n,q) \notin \{(3,2^2),(3,2^3),(3,3^2),(5,2^2)\}$. Here a power of y^2 has type $(n-1)^-_{q_0} \perp I_1$, so

$$|C_{\mathrm{PGL}_n(q_0)}(y^2)| = q_0^{n-1}-1.$$

By applying Lemma 3.4.1(ii)(b) in the usual way, we see that $t\theta$ is contained in a unique reducible maximal subgroup, and by Bray et al. [8], there are at most $2+\delta_{n,5}+\log\log q$ classes of irreducible maximal subgroups, so by (4.1) and Theorem 4.2.4,

$$P(x,t\theta) < \frac{4}{3q} + (2+\delta_{n,5}+\log\log q)\cdot(q_0^{n-1}-1)\cdot\frac{4}{3q^{n-1}} < \frac{1}{2}.$$

For the remainder of the proof we may assume that n is even. For now assume that $n \geqslant 12$. If we are in case (i), then

$$|C_{\mathrm{PGL}_n(q_0)}(y^2)| = q_0^{\delta_{q_0,3}}(q_0+1)(q_0^{n-2}-1)$$

and, by Proposition 6.4.10,

$$P(x,t\theta) < \delta_{q_0,3}\left(\frac{3}{q^2}+\frac{6}{q^{n-1}}+\frac{6}{q^{n-4}}\right)+\frac{1}{q^4}+\frac{2}{q^{n-4}}+\frac{1}{q^{n-2}}+\frac{1}{q^{n-1}}$$
$$+(d(n)+\log\log q+4-3\delta_{2,p})q_0^{\delta_{q_0,3}}(q_0^{n-2}-1)(q_0+1)\frac{2}{q^{n-3+2/n}} < \frac{1}{2}.$$

Now consider case (ii). Write $N = |C_{\mathrm{PGL}_n(q_0)}(y^2)|$. Then

$$N \leqslant \begin{cases} q_0(q_0+1)(q_0^{n-2}-1) & \text{if } \eta = + \\ q_0(q_0+1)(q_0^{n/2-1}-1)^2 & \text{if } \eta = - \text{ and } n \equiv 0 \pmod 4 \\ q_0(q_0+1)(q_0^{n/2+1}-1)(q_0^{n/2-3}-1) & \text{if } \eta = - \text{ and } n \equiv 2 \pmod 4). \end{cases}$$

Therefore,

$$P(x,t\theta) < \frac{4}{q^2}+\frac{1}{q^4}+\frac{4}{q^{n/2-1/2}}+\frac{1}{q^{n/2+1/2}}+\frac{1}{q^{n/2+3}}+\frac{1}{q^{n-6}}+\frac{6}{q^{n-4}}+\frac{13}{q^{n-2}}$$
$$+(d(n)+\log\log q+4)\cdot N\cdot\frac{2}{q^{n-3+2/n}} < \frac{1}{2}.$$

It remains to assume that $n \in \{4,6,8,10\}$. For now assume that $n \geqslant 6$. Now y^2 has type $(n)^{\pm}_{q_0}$ and

$$|C_{\mathrm{PGL}_n(q_0)}(y^2)| \leqslant \tfrac{q_0^n-1}{q_0-1}.$$

Since y^2 is not contained in any $\mathcal{C}_1$ subgroups of $\mathrm{PGL}_n(q_0)$, by Lemma 3.4.1(ii)(b) implies that $t\theta$ is not contained in any $\mathcal{C}_1$ subgroups of G. Consulting [8], we see that G has at most 4 classes of $\mathcal{C}_2$ subgroups, at most $k-2\delta_{2,p}+\log\log q$ further classes of irreducible maximal subgroups, and together Proposition 4.2.3 and Theorem 4.2.4 establish $\mathrm{fpr}(x, G/H) \leqslant f(q)$ for all $H \in \mathcal{M}(G,t\theta)$, where

$$k = \begin{cases} 7 \text{ if } n=10 \\ 3 \text{ if } n=8 \\ 5 \text{ if } n=6 \end{cases} \quad \text{and} \quad f(q) = \begin{cases} 2q^{-7.2} & \text{if } n=10 \\ 2q^{-5.25} & \text{if } n=8 \\ (q^4-q^3+q^2-q+1)^{-1} & \text{if } n=6. \end{cases}$$

Therefore, if $e \geqslant 3$, then

$$P(x,t\theta) \leqslant (4+k-2\delta_{2,p}+\log\log q)\cdot\tfrac{q_0^n-1}{q_0-1}\cdot f(q) < \frac{1}{2}.$$

Now assume that $e = 2$. Let y' be a power of y whose order is a primitive prime divisor of $q^{n/2}-1$. Since $\frac{n}{2} \not\equiv 2 \pmod 4$, by [20, Proposition 3.3.2], y' centralises the decomposition $\mathbb{F}_{q^2}^n = U \oplus U^*$ where U is a totally singular $\frac{n}{2}$-space on which y' acts irreducibly. Therefore, y' is contained in a unique subgroup of type $\mathrm{GL}_{\frac{n}{2}}(q^2)$ and no further $\mathcal{C}_2$ subgroups. Therefore, we obtain

$$P(x, t\theta) \leqslant \left(1 + (k - 2\delta_{2,p} + \log\log q) \cdot \tfrac{q_0^n - 1}{q_0 - 1}\right) \cdot f(q) < \frac{1}{2}.$$

Finally assume that $n = 4$. By Proposition 6.2.9, we will assume that $q \notin \{2^2, 2^3, 3^2\}$. Now y has type $(4)_{q_0}^-$, so $|C_{\mathrm{PGL}_4(q_0)}(y^2)| = (q_0 + 1)(q_0^2 + 1)$. By Lemma 3.4.1(ii)(b), $t\theta$ is not contained in any reducible maximal subgroup of G since y^2 is not contained in any reducible maximal subgroups of $\mathrm{PGL}_4(q_0)$. From [8], there are at most $5 + \log\log q$ classes of irreducible maximal subgroups. If $e \geqslant 3$, then, by Theorem 4.2.4,

$$P(x, t\theta) \leqslant \frac{(4 + \log\log q) \cdot (q_0^2 - q_0 + 1)}{q^2 - q + 1} + \frac{(2, q+1)(q_0^2 - q_0 + 1)}{q} < \frac{1}{2}.$$

If $e = 2$, then let k be 2 if $q_0 \equiv 3 \pmod 4$ and 1 otherwise, so by Proposition 6.4.11,

$$P(x, t\theta) \leqslant \frac{(4 + \log\log q) \cdot (q_0^2 - q_0 + 1)}{q^2 - q + 1} + \frac{k(q_0 + 1) \cdot (2, q+1)(q^4 + 1)}{q^5 + q^2} < \frac{1}{2}.$$

In all cases $P(x, \theta) \to 0$ as $q \to \infty$. □

6.5 Case II: Linear Automorphisms

Now $G = \langle T, \theta \rangle$ for $T = \mathrm{PSU}_n(q)$ with $n \geqslant 3$ and $\theta \in \langle \mathrm{PGU}_n(q), \gamma \rangle$. Recall

(a) $G \leqslant \mathrm{PGU}_n(q)$
(b) $G \not\leqslant \mathrm{PGU}_n(q)$.

Cases II(a) and II(b) will be considered in Sects. 6.5.1 and 6.5.2, respectively.

6.5.1 Case II(a)

Let $T = \mathrm{PSU}_n(q)$ and let $G = \langle T, \theta \rangle \in \mathcal{A}_-$ in Case II(a). Therefore, $G = T$ or $(n, q+1) > 1$ and we write $G = \langle T, \delta^\ell \rangle$ for some $0 < \ell < (n, q+1)$. As in Sect. 5.5.1, in the following proof, we closely follow [11, Sections 5.10 and 5.11].

Proposition 6.5.1 *Let $G = \langle T, \theta \rangle \in \mathcal{A}_-$. In Case II(a), $u(G) \geqslant 2$ and as $q \to \infty$ we have $u(G) \to \infty$.*

Proof If $G = T$, then the result follows from [11, Propositions 5.20 and 5.21]. Therefore, for the remainder, we may assume that $(n, q+1) > 1$ and write $G = \langle T, \delta^\ell \rangle$ for some fixed $0 < \ell < (n, q+1)$. Let $s = y^\ell$ where y has type $[n]^-$ if n is odd and $[n-1]^- \perp I_1$ if n is even. Note that $s \in T\delta^\ell$, since $\det(s) = \alpha^\ell$. By Proposition 6.2.9, we may assume that $q \geqslant 11$ if $n \in \{3, 4\}$, $q \geqslant 4$ if $n \in \{5, 6, 7, 8\}$.

If n is odd, then s acts irreducibly on $V = \mathbb{F}_{q^2}^n$, and if n is even, then, by Lemma 2.3.3, s is contained in a unique reducible maximal subgroup of G, of type $\mathrm{GU}_1(q) \perp \mathrm{GU}_{n-1}(q)$. Now let $H \in \mathcal{M}(G, s)$ be irreducible. Now $\Phi_k^*(q^2)$ divides $|s|$ and Lemma 2.6.4 implies $\Phi_k^*(q^2) > 2k+1$, so applying Theorem 2.6.3, we see that H is a subfield or field extension subgroup. In the former case, $|s|$ does not divide $|H|$, and in the latter, the degree of the field extension divides (n, k), so n must be odd and H has type $\mathrm{GU}_{n/k}(q^k)$ for some prime k dividing n and [11, Lemma 2.12] implies s is contained in a unique subgroup of type $\mathrm{GU}_{n/k}(q^k)$ for each possible k.

Let $x \in G$ have prime order. For now assume that $n \geqslant 4$ is even. Then $\mathcal{M}(G, s) = \{H\}$ for H of type $\mathrm{GU}_1(q) \times \mathrm{GU}_{n-1}(q)$. If $n \geqslant 6$, by Theorem 4.1.1,

$$P(x, s) < \frac{1}{q^2} + \frac{2}{q^{n-4}} + \frac{1}{q^{n-2}} + \frac{1}{q^{n-1}} < \frac{1}{2}$$

(recall $q \geqslant 4$ if $n = 6$), and by (4.1), $P(x, s) < \frac{4}{3q} < \frac{1}{2}$ if $n \geqslant 4$ (recall $q \geqslant 4$).

Now assume $n \geqslant 3$ is odd. Then $\mathcal{M}(G, s) = \{H_k \mid k \text{ is a prime divisor of } n\}$, where H_k has type $\mathrm{GU}_{n/k}(q^k)$. By Proposition 4.2.3, if $n \geqslant 9$, then

$$P(x, s) < (n-2) \cdot \frac{2}{q^{n-3}} < \frac{1}{2}$$

and $P(x, s) < \frac{2}{q^{n-3}} \leqslant \frac{1}{2}$ if $n \in \{5, 7\}$. If $n = 3$, then $q \geqslant 11$ and Theorem 4.2.4 implies that $P(x, s) \leqslant (q^2 - q + 1)^{-1} < \frac{1}{2}$. In all cases, $P(x, s) \to 0$ as $q \to \infty$. □

6.5.2 Case II(b)

This section completes the proof of Theorems 6.1 and 6.2, by considering Case II(b). Here $G = \langle T, \theta \rangle$ where θ is either γ or $\delta_2\gamma$ (for even $(n, q+1)$ in the latter case).

To avoid repetition, we refer to some tables in Sect. 6.4.2, with the convention that $i = f$, so $e = 1$, $\varphi^i = \gamma$ and $q_0 = q$. By (6.5) and Remark 6.2.5(iv), the centraliser $C_{\mathrm{PGU}_n(q)}(\theta)$ is given in the Z_σ column of Fig. 6.4. Let $t \in C_{\mathrm{PSU}_n(q)}(\theta)$ be the element y in Fig. 6.5.

	type of H	$m(H)$	conditions
$\mathcal{C}_1$	P_1	1	n odd, q odd
	P_2	1	n odd, q odd
	$P_{\frac{n}{2}-1}$	2	$\theta = \delta_2\gamma$, $n \equiv 0 \pmod 4$
	$\mathrm{GU}_1(q) \times \mathrm{GU}_{n-1}(q)$	1	n odd, q even
		4	n even, $q = 3$
	$\mathrm{GU}_2(q) \times \mathrm{GU}_{n-2}(q)$	1	n even or q even
	$\mathrm{GU}_3(q) \times \mathrm{GU}_{n-3}(q)$	1	n odd
	$\mathrm{GU}_{\frac{n}{2}-3}(q) \times \mathrm{GU}_{\frac{n}{2}+3}(q)$	1	$\theta = \delta_2\gamma$, $n \equiv 2 \pmod 4$, $q \equiv 3 \pmod 4$
	$\mathrm{GU}_{\frac{n}{2}-2}(q) \times \mathrm{GU}_{\frac{n}{2}+2}(q)$	4	$\theta = \delta_2\gamma$, $n \equiv 2 \pmod 4$, $q = 3$
	$\mathrm{GU}_{\frac{n}{2}-1}(q) \times \mathrm{GU}_{\frac{n}{2}+1}(q)$	1	$\theta = \delta_2\gamma$, $n \equiv 2 \pmod 4$, $q \equiv 3 \pmod 4$
$\mathcal{C}_2$	$\mathrm{GL}_{\frac{n}{2}}(q^2)$	$q+1$	n even
	$\mathrm{GU}_{\frac{n}{2}}(q) \wr S_2$	$q+1$	n even
	$\mathrm{GU}_3(q) \wr S_3$	1	$n = 9$
$\mathcal{C}_5$	$\mathrm{O}_n(q)$	M	n odd, q odd
	$\mathrm{Sp}_n(q)$	M	n even
	$\mathrm{O}^+_n(q)$	M	$\theta = \gamma$, n even, q odd
	$\mathrm{O}^\eta_n(q)$	M	$\theta = \delta_2\gamma$, n even, q odd
	$\mathrm{GU}_9(q^{1/3})$	M	$n = 9$ and $3 \mid f$

Note: $M = |C_{\mathrm{Inndiag}(\mathrm{soc}(G))}(t^2) : C_{\mathrm{Inndiag}(\mathrm{soc}(H))}(t^2)|$

Fig. 6.7 Case II(b): Description of $\mathcal{M}(G, t\theta)$

Proposition 6.5.2 *Assume that $n \geqslant 7$ is odd or $n \geqslant 12$ is even. Then the maximal subgroups of G that contain $t\theta$ are listed in Fig. 6.7, where $m(H)$ is an upper bound on the multiplicity of the subgroups of type H in $\mathcal{M}(G, t\theta)$.*

Proof Let $H \in \mathcal{M}(G, t\theta)$. A power of t^2 has type $J_3 \perp I_{n-3}$ or $(2)^-_q \perp I_{n-2}$ unless n is even and $q = 3$, when it has type $-I_2 \perp I_{n-2}$. In particular, a power of t^2 is an element z satisfying $\nu(z) = 2$, so by Proposition 4.2.5, one of the following holds

(i) $H \in \mathcal{C}_1 \cup \mathcal{C}_2 \cup \mathcal{C}_5$
(ii) $H \in \mathcal{C}_4$ has type $\mathrm{GU}_2(q) \otimes \mathrm{GU}_{n/2}(q)$ with $q = 3$ and n even
(iii) $H \in \mathcal{S}$ has socle $\mathrm{PSU}_3(3)$ with $n = 7$ and $q = p \equiv 2 \pmod 3$ odd.

We begin by eliminating the possibilities in (ii) and (iii). For (ii), let $n = 2m$ be even, $q = 3$ and H have type $\mathrm{GU}_2(q) \otimes \mathrm{GU}_m(q)$. Write $t^2 = -I_2 \perp x$ and suppose $g \otimes h = t^2$. Then $-1 = \lambda\mu$ for some eigenvalues λ and μ of g and h, respectively. Therefore, $\mu = -\lambda \in \mathbb{F}_{3^2}^\times$, contradicting no eigenvalue of x being in $\mathbb{F}_{3^2}^\times$.

For (iii), let $\mathrm{PSU}_3(3) \leqslant H \leqslant \mathrm{Aut}(\mathrm{PSU}_3(3))$. The only prime divisors of $|H|$ are 2, 3 and 7, but $|t^2|$ is divisible by a primitive prime divisor of $q^4 - 1$, which satisfies $r \equiv 1 \pmod 4$, so $t^2 \notin H$. Therefore, $H \in \mathcal{C}_1 \cup \mathcal{C}_2 \cup \mathcal{C}_5$.

First assume that $H \in \mathcal{C}_1$. We will apply Shintani descent. Let X be the simple algebraic group $\mathrm{PGL}_n(\overline{\mathbb{F}}_p)$ and let σ be the Frobenius endomorphism φ^f. Notice that $t\theta \in \mathrm{PGU}_n(q)\varphi^f \subseteq \mathrm{PGL}_n(q^2)\varphi^f = X_{\sigma^2}\sigma$. Moreover, $H \cap \mathrm{PGU}_n(q) \leqslant Y_{\sigma^2}$ for a closed connected σ-stable subgroup Y of X. By Lemma 3.3.2, the $\langle X_{\sigma^2}, \widetilde{\sigma}\rangle$-conjugates of Y_{σ^2} that are normalised by $t\theta$ correspond to the X_σ-conjugates of Y_σ that contain t^2. It is easy to determine the maximal reducible overgroups of t^2 in $\mathrm{PGL}_n(q)$ and these give the maximal reducible overgroups of $t\theta$ in G that feature in Fig. 6.7 (see the proof of Proposition 6.4.10 for further details).

Next assume that $H \in \mathcal{C}_2$. Write $H = N_G(H_0)$ where $H_0 = H \cap T$ is the stabiliser in T of a direct sum decomposition $\mathbb{F}_{q^2}^n = U_1 \oplus \cdots \oplus U_k$ where $\dim U_i = n/k$ and $k > 1$. Let B be the centraliser in H_0 of this decomposition. For now assume that $\theta = \gamma$ or $\eta = +$. Let m be 2 if n is even and 3 if n is odd. Then we may fix a suitable power z of t^2 of type $(n-m)_q^- \perp I_m$. A power z' of z has order a primitive prime divisor r of $q^{n-m}-1$. Since $r \geqslant n-m+1 > k$, we see that $z' \in B$. However, $z' = z_1 \oplus z_2 \oplus I_m$ with respect to a decomposition $\mathbb{F}_{q^2}^n = Z_1 \oplus Z_2 \oplus Z$, where $\dim Z_i = (n-m)/2$ and $\dim Z = m$, and z_i acts irreducibly on Z_i. This implies that n is even and $k = 2$ (and t^2 is contained in at most $q+1$ subgroups of a given type) or $n = 9$ and $k = 3$ (and t^2 is contained in a unique such subgroup).

We may now assume that $\theta = \delta_2\gamma$ and $\eta = -$. First assume that $n \equiv 0 \pmod 4$. Let z be a power of t^2 of type $(n-2)_q^+ \perp I_2$ of order a primitive prime divisor r of $q^{n/2-1}-1$. Now $r \equiv 1 \pmod{(n/2-1)}$, but $r \neq n/2$ since r is odd, so $r = n-1 \neq k$ or $r \geqslant 3n/2-2 > n \geqslant k$. Therefore, $z \in B$. Now $z = z_1 \oplus z_2 \oplus I_2$ with respect to a decomposition $\mathbb{F}_{q^2}^n = Z_1 \oplus Z_2 \oplus Z$, where $\dim Z_i = (n-2)/2$ and $\dim Z = 2$, and z_i acts irreducibly on Z_i. As above, this implies that $k = 2$ and t^2 is contained in at most $q+1$ subgroups of a given type. Now assume that $n \equiv 2 \pmod 4$. Let z be a power of t^2 of type $\left(\frac{n}{2}+1\right)_q^- \perp \left(\frac{n}{2}-3\right)^- \perp I_2$ of order rs where r is a primitive prime divisor of $q^{n/2+1}-1$ and s is a primitive prime divisor of $q^{n/2-3}-1$. Now $r \equiv 1 \pmod{(n/2+1)}$ and $s \equiv 1 \pmod{(n/2-3)}$ and as in the previous case we deduce that $r \neq k$ and $s \neq k$, so $z \in B$. Now $z = z_1^1 \oplus z_1^2 \oplus z_2 \oplus z_2^2 \oplus I_2$ with respect to a decomposition $\mathbb{F}_{q^2}^n = Z_1^1 \oplus Z_1^2 \oplus Z_2^1 \oplus Z_2^2 \oplus Z$, where $\dim Z_1^j = (n+2)/4$, $\dim Z_2^j = (n-6)/4$ and z_i^j acts irreducibly on Z_i^j. This implies that $k = 2$ and t^2 is contained in at most $2(q+1)$ subgroups of a given type.

Finally assume that $H \in \mathcal{C}_5$. We postpone subgroups of type $\mathrm{GU}_n(q^{1/k})$, so we may assume that n is even or q is odd. If n and q are even, then H has type $\mathrm{Sp}_n(q)$; if n and q are odd, then H has type $\mathrm{O}_n(q)$; and if n is even and q is odd, then H has type $\mathrm{Sp}_n(q)$ or $\mathrm{O}_n^\upsilon(q)$, where $\upsilon = +$ if $\theta = \gamma$ or e is even and $\upsilon = \eta$ if $\theta = \delta_2\gamma$.

Write $H_0 = H \cap T$, so $|G : T| = |H : H_0| = 2$. Let c be the number of G-classes of subgroups of G of type H. From [50, Propositions 4.5.5 and 4.5.6], we see that

$$c = \frac{|\operatorname{Inndiag}(T) : T|}{|\operatorname{Inndiag}(\operatorname{soc}(H_0)) : H_0|}.$$

From the description of the conjugacy classes of elements of prime order in [20, Chapter 3], we see that $(t^2)^T = (t^2)^{\mathrm{Inndiag}(T)}$ and $(t^2)^{H_0} = (t^2)^{\mathrm{Inndiag}(H_0)}$, so

$$\frac{|C_{\mathrm{Inndiag}(T)}(t^2) : C_T(t^2)|}{|C_{\mathrm{Inndiag}(\mathrm{soc}(H_0))}(t^2) : C_{H_0}(t^2)|} = \frac{|\,\mathrm{Inndiag}(T) : T|}{|\,\mathrm{Inndiag}(\mathrm{soc}(H_0)) : H_0|}.$$

Moreover, $(t^2)^T \cap H_0 = (t^2)^{H_0}$. Therefore, the number of subgroups of G of type H that contain t^2 is

$$c \cdot \frac{|G|}{|H|}\frac{|(t^2)^G \cap H|}{|(t^2)^G|} = c \cdot \frac{|T|}{|H_0|}\frac{|(t^2)^{H_0}|}{|(t^2)^T|} = c \cdot \frac{|C_T(t^2)|}{|C_{H_0}(t^2)|} = \frac{|C_{\mathrm{Inndiag}(T)}(t^2)|}{|C_{\mathrm{Inndiag}(\mathrm{soc}(H_0))}(t^2)|}.$$

It remains to assume that H has type $\mathrm{GU}_n(q^{1/k})$ for an odd prime divisor k of f (recall that $q = p^f$). In order for $|t^2|$ to divide $|\,\mathrm{GU}_n(q^{1/k})|$ we must have $(n,k) = (9,3)$, and arguing as in the previous case we see that t^2 is contained in $|C_{\mathrm{Inndiag}(T)}(t^2) : C_{\mathrm{Inndiag}(\mathrm{soc}(H_0))}(t^2)|$ subgroups of this type. □

Proposition 6.5.3 *Let $G = \langle T, \theta \rangle \in \mathcal{A}$ where $T = \mathrm{PSU}_n(q)$. In Case II(b), $u(G) \geqslant 2$ and as $q \to \infty$ we have $u(G) \to \infty$.*

Proof Let $x \in G$ have prime order. We begin by computing the parameter M that features in Fig. 6.7. If $n \geqslant 7$ is odd and q is odd, then t^2 has type $(n-3)^- \perp J_3$ and

$$\frac{|C_{\mathrm{PGU}_n(q)}(t^2)|}{|C_{\mathrm{PSO}_n(q)}(t^2)|} = \frac{q^2(q^{(n-3)/2}+1)(q^{(n-3)/2} - (-)^{(n-3)/2})}{q(q^{(n-3)/2}+1)} \leqslant q^{(n-1)/2} + q.$$

Similarly, if $n \geqslant 12$ is even and either $\theta = \gamma$ or $\theta = \delta_2\gamma$ and $\eta = +$, then t^2 has type $(n-2)^- \perp (2)^-$ and

$$\frac{|C_{\mathrm{PGU}_n(q)}(t^2)|}{|C_{\mathrm{PGSp}_n(q)}(t^2)|} = \frac{|C_{\mathrm{PGU}_n(q)}(t^2)|}{|C_{\mathrm{PDO}^+_n(q)}(t^2)|} \leqslant q^{(n-2)/2} + 1,$$

if $\theta = \delta_2\gamma$, $\eta = -$ and $n \equiv 0 \pmod 4$, then

$$\frac{|C_{\mathrm{PGU}_n(q)}(t^2)|}{|C_{\mathrm{PGSp}_n(q)}(t^2)|} = \frac{|C_{\mathrm{PGU}_n(q)}(t^2)|}{|C_{\mathrm{PDO}^-_n(q)}(t^2)|} \leqslant q^{(n-2)/2} + 1,$$

and if $\theta = \delta_2\gamma$, $\eta = -$ and $n \equiv 2 \pmod 4$, then

$$\frac{|C_{\mathrm{PGU}_n(q)}(t^2)|}{|C_{\mathrm{PGSp}_n(q)}(t^2)|} = \frac{|C_{\mathrm{PGU}_n(q)}(t^2)|}{|C_{\mathrm{PDO}^-_n(q)}(t^2)|} \leqslant (q^{(n+2)/4} + 1)(q^{(n-6)/4} + 1).$$

Finally, if $n = 9$ and 3 divides f, then t^2 either has type $(6)^- \perp (2)^- \perp I_1$ or $6^- \perp J_3$, but in either case

$$\frac{|C_{\mathrm{PGU}_9(q)}(t^2)|}{|C_{\mathrm{PGU}_9(q^{1/3})}(t^2)|} \leqslant \frac{(q^6-1)(q+1)^2}{(q^2-1)(q^{1/3}+1)^2} \leqslant q^{4/3}(q^4+q^2+1).$$

First assume that $n \geqslant 7$ is odd. By Proposition 6.2.9, we will assume that $(n, q) \neq (7, 2)$. Let $\beta = 1$ if $n = 9$ and 3 divides f. If q is even, then

$$P(x, t\theta) < \frac{1}{q^2} + \frac{1}{q^4} + \frac{1}{q^6} + \frac{1}{q^{n-3}} + \frac{7}{q^{n-2}} + \frac{1}{q^{n-1}} + \frac{3}{q^{n+1}} + (\delta_{n,9} + \beta q^{4/3}(q^4+q^2+1)) \cdot \frac{2}{q^6} < \frac{1}{2},$$

and if q is odd, then

$$P(x, t\theta) < \frac{1}{q^2} + \frac{1}{q^4} + \frac{1}{q^6} + \frac{1}{q^{n-3}} + \frac{2}{q^{n-2}} + \frac{4}{q^{n-4}} + \frac{2}{q^{n/2}} + \frac{1}{q^{n+1}} + (\delta_{n,9} + \beta q^{4/3}(q^4+q^2+1) + q^{(n-1)/2} + q) \cdot \frac{2}{q^{n-3+2/n}} < \frac{1}{2}.$$

Next consider $n = 5$, where, by Proposition 6.2.9, we will assume that $q \geqslant 4$. Here t^2 has type $(4)_q^- \perp I_1$, which has order $r = \Phi_4^*(q) \geqslant 13$ that satisfies $r \equiv 1 \pmod 4$ by Lemma 2.6.4. Arguing as in the proof of Proposition 6.5.2, via Lemma 3.3.2, $t\theta$ is contained in a $\mathrm{GU}_1(q) \times \mathrm{GU}_4(q)$ subgroup and no further reducible subgroups. Since t^{10} has order divisible by a primitive prime divisor of $q^4 - 1$ and acts irreducibly on two nonisomorphic 2-spaces of $\mathbb{F}_{q^2}^5$, inspecting [8, Tables 8.20 and 8.21], all irreducible maximal subgroups of G that contain t^2 have type $\mathrm{SO}_5(q)$. Arguing as in the proof of Proposition 6.5.2, we see the number of subgroups of type $\mathrm{SO}_5(q)$ that contain t^2 is $|C_{\mathrm{PGU}_5(q)}(t^2) : C_{\mathrm{PSO}_5(q)}(t^2)| = q^2 - 1$. Therefore,

$$P(x, t\theta) < \frac{4}{3q} + (q^2-1) \cdot \frac{4}{3q^4} < \frac{1}{2}.$$

Now consider $n = 3$, where, by Proposition 6.2.9, we will assume that $q \geqslant 11$. Here $t^2 = g \perp I_1$ where $|g| = (q+1)/(q+1, 2) > 2$. Therefore, $t\theta$ is contained in a $\mathrm{GU}_1(q) \times \mathrm{GU}_2(q)$ subgroup and no further reducible subgroups. Arguing as in the proof of Proposition 6.5.2, we see the number of subgroups of type $\mathrm{SO}_3(q)$ that contain t^2 is $|C_{\mathrm{PGU}_3(q)}(t^2) : C_{\mathrm{PSO}_3(q)}(t^2)| = q + 1$. Since $q \geqslant 13$, for $k \in \{2, 3\}$ we have $|g|/(|g|, k) > 2$, so $t^{2k} = [\lambda, \lambda^q, 1]$ where $\lambda \in \mathbb{F}_{q^2}^\times \setminus \mathbb{F}_q^\times$. Therefore, g is contained in at most one subgroup of type $\mathrm{GU}_1(q) \wr S_3$ and no subgroups of type $\mathrm{GU}_1(q^3)$. Consulting, [8, Tables 8.20 and 8.21], all remaining maximal subgroups of G do not contain elements of order $|t^2|$. Therefore,

$$P(x, t\theta) < \frac{4}{3q} + \frac{q+1}{q^2-q+1} < \frac{1}{2}.$$

For the remainder of the proof, we may assume that n is even. For now assume that $n \geqslant 12$. If $\theta = \gamma$, then

$$P(x, t\theta) < \frac{4\delta_{q,3}}{q^2} + \frac{1}{q^4} + \frac{2+8\delta_{3,q}}{q^{n-4}} + \frac{1+4\delta_{3,q}}{q^{n-2}} + \frac{1+4\delta_{3,q}}{q^{n-1}} + (2q+2+(2-\delta_{2,p})(q^{(n-2)/2}+1)) \cdot \frac{2}{q^{n-3}} < \frac{1}{2},$$

and if $\theta = \delta\varphi^i$, then

$$P(x, t\theta) < \frac{4}{q^2} + \frac{1}{q^4} + \frac{1}{q^{n-6}} + \frac{26}{q^{n-4}} + \frac{2}{q^{n-2}} + \frac{15}{q^{n-1}} + \frac{6}{q^{(n+2)/2}} + (2q+2+2(q^{(n+2)/4}+1)(q^{(n-6)/4}+1)) \cdot \frac{2}{q^{n-3}} < \frac{1}{2}.$$

We now handle the remaining cases where n is even. First assume that $n = 10$, so t^2 has type $(10)^\eta$ (with the convention that $\eta = -$ if $\theta = \gamma$). Arguing as in the proof of Proposition 6.5.2, $t\theta$ is not contained in any reducible maximal subgroups. The order of t^2 is $r \in \{\Phi^*_{10}(q), \Phi^*_5(q)\}$, so Lemma 2.6.4 implies that all prime divisors of r are 1 (mod 5) and $r \geqslant 16$, so, in fact, $r \geqslant 31$. Therefore, inspecting [8, Tables 8.62 and 8.63], the only possible types of irreducible maximal subgroup of G that could contain t^2 are those of type $\mathrm{GU}_2(q^5)$, $\mathrm{GL}_5(q^2)$, $\mathrm{GU}_5(q) \wr S_2$, $\mathrm{Sp}_{10}(q)$ and $\mathrm{SO}^\eta_{10}(q)$.

The number of subgroups of types $\mathrm{Sp}_{10}(q)$ and $\mathrm{SO}^\eta_{10}(q)$ that contain t^2 is

$$M = \frac{|C_{\mathrm{PGU}_{10}(q)}(t^2)|}{|C_{\mathrm{PGSp}_{10}(q)}(t^2)|} = \frac{|C_{\mathrm{PGU}_n(q)}(t^2)|}{|C_{\mathrm{PDO}^\eta_n(q)}(t^2)|} = \frac{q^5+1}{q+1}.$$

Now let us determine the multiplicities of $\mathcal{C}_2$ and $\mathcal{C}_3$ subgroups. First assume that $H \in \mathcal{C}_2$. Since t^2 has odd order, if $t^2 \in H$ then t^2 centralises the decomposition $\mathbb{F}_{q^2}^{10} = U_1 \oplus U_2$ where $\dim U_1 = \dim U_2 = 5$. If $\eta = +$, then $\{U_1, U_2\}$ must be a dual pair of totally singular subspaces and t^2 centralises a unique such decomposition, so t^2 is contained in a unique subgroup of type $\mathrm{GL}_5(q^2)$ and no subgroups of type $\mathrm{GU}_5(q) \wr S_2$. If $\eta = -$, then U_1 and U_2 must be orthogonal nondegenerate subspaces and again t^2 centralises a unique such decomposition, so t^2 is contained in a unique subgroup of type $\mathrm{GU}_5(q) \wr S_2$ and no subgroups of type $\mathrm{GL}_5(q^2)$.

Now assume that H has type $\mathrm{GU}_2(q^5)$. Write $H \cap T = H_0 = B.5$ and let π be the field extension embedding. Now $t^2 \in B$ as all prime divisors of $r = |t^2|$ are at least 11. Let $b \in B$ satisfy $\pi(b) = t^2$. Write $\Lambda = \{\lambda, \lambda^{q^2}, \lambda^{q^4}, \lambda^{q^6}, \lambda^{q^8}\}$ where $|\lambda| = r$. For now assume that $\eta = +$, so $t^2 = [\Lambda, \Lambda^{-1}]$. Then $b = [\lambda^{q^i}, \lambda^{-q^i}]$ where $0 \leqslant i \leqslant 4$, so there are 5 possibilities for b up to B-conjugacy and consequently 1 possibility up to H_0-conjugacy. Therefore, $|(t^2)^T \cap H_0| = |b^{H_0}|$. In addition, $|C_{\mathrm{GU}_{10}(q)}(t^2)| = q^{10} - 1 = |C_{\mathrm{GU}_2(q^5)}(b)|$, so t^2 is contained in a unique subgroup of type $\mathrm{GU}_2(q^5)$. If $\eta = -$, then $t^2 = [\Lambda, \Lambda^q]$, so $b = [\lambda^{q^i}, (\lambda^q)^{q^j}]$ where $0 \leqslant i, j \leqslant 4$. There are 25 possibilities for b up to B-conjugacy and 5 up to H_0-conjugacy, so as before we deduce that t^2 is contained in 5 subgroups of G of this type.

Therefore,

$$P(x, t\theta) < \left(1 + 5 + 2 \cdot \frac{q^5 + 1}{q + 1}\right) \cdot \frac{2}{q^7} < \frac{1}{2}.$$

The cases $n \in \{6, 8\}$ are very similar. In both cases, by Proposition 6.2.9 we assume that $q \geqslant 4$. If $n = 8$, then t^2 has type $(8)^-$ of order $r = \Phi_8^*(q) \geqslant 25$ and all prime divisors of r are 1 (mod 8), and our usual arguments allow us to conclude that the maximal subgroups of G containing t^2 are one of type $\mathrm{GL}_4(q^2)$ and $(q+1)(q^2+1)$ of types $\mathrm{Sp}_8(q)$ and $\mathrm{SO}_8^-(q)$ (where q is odd in the latter case), so we obtain

$$P(x, t\theta) < (1 + 2(q + 1)(q^2 + 1)) \cdot \frac{2}{q^5} < \frac{1}{2}.$$

If $n = 6$, then t^2 has type $(6)^\eta$ (again, with the convention that $\eta = -$ if $\theta = \gamma$) and we deduce that the maximal subgroups of G that contain t^2 are a unique subgroup of type $\mathrm{GL}_3(q^2)$ if $\eta = +$ and of type $\mathrm{GU}_3(q) \wr S_2$ if $\eta = -$ and $q^2 - q + 1$ subgroups of types $\mathrm{Sp}_6(q)$ and $\mathrm{SO}_6^\eta(q)$ (q odd), so

$$P(x, t\theta) \leqslant \frac{q^2 - q + 3}{q^4 - q^3 + q^2 - q + 1} < \frac{1}{2}.$$

Finally assume that $n = 4$. By Proposition 6.2.9, we can assume that $q \geqslant 11$. Since t^2 has type $(4)^-$, our usual application of Lemma 3.4.1(ii)(b) implies that $t\theta$ is not contained in any reducible maximal overgroups. The order r of t^2 satisfies $r = \Phi_4^*(q) \geqslant 13$ and all prime divisors of r are 1 (mod 5). Therefore, consulting the list of maximal subgroups of G in [8, Tables 8.10 and 8.11], we see that the only types of maximal subgroup that could contain t^2 are $\mathrm{GL}_2(q^2)$, $\mathrm{Sp}_4(q)$ and, if q is odd, $\mathrm{SO}_4^-(q)$. Arguing as in the previous cases, t^2 stabilises a unique decomposition $\mathbb{F}_{q^2}^4 = U \oplus U^*$ where U is a maximal totally singular subspace, so $t\theta$ is contained in at most one subgroup of type $\mathrm{GL}_2(q^2)$. Moreover, $t\theta$ is contained in at most $|C_{\mathrm{PGU}_4(q)}(t^2) : C_{\mathrm{PSO}_4^-(q)}(t^2)| = q - 1$ subgroups of type $\mathrm{SO}_4^-(q)$. It remains to estimate the number m of subgroups of type $\mathrm{Sp}_4(q)$ that contain $t\theta$. There is a unique $\widetilde{G}$-class of such subgroups, so $m = \sum_{i=1}^k |C_{\widetilde{G}}(t_i\theta) : C_{\widetilde{H}}(t_i\theta)|$ where $\widetilde{G} =$

$\langle \mathrm{PGU}_4(q), \gamma\rangle$ and $\widetilde{H} = N_{\widetilde{G}}(\widetilde{H}) = C_{\widetilde{G}}(\gamma)$, and where $(t\theta)^{\widetilde{G}} \cap H = \bigcup_{i=1}^{k} (t_i\theta)^{\widetilde{H}}$. If g centralises $t\theta$, then g centralises the power γ, so $C_{\widetilde{G}}(t_i\theta) = C_{\widetilde{H}}(t_i\theta)$. If $t\theta$ is $\widetilde{G}$-conjugate to $s\theta$, then t^2 and s^2 have the same eigenvalues, so as we argued in the proof of Proposition 6.4.11, t is $\widetilde{H}$-conjugate to s if $q \not\equiv 3 \pmod 4$ and there are at most two choices for t up to $\widetilde{H}$-conjugacy if $q \equiv 3 \pmod 4$. Therefore, $m = k \leqslant (2, q+1)$. Now using the fixed point ratio bounds in Theorem 4.2.4 we conclude that

$$P(x, t\theta) \leqslant \frac{q+1}{q^2-q+1} + \frac{(2,q+1)^2(q^4+1)}{q^5+q^2} < \frac{1}{2}$$

In every case, $P(x, t\theta) \to 0$ as $q \to \infty$. This completes the proof. □

Combining Propositions 6.4.5, 6.4.12, 6.5.1, 6.5.3 yields Theorems 6.1 and 6.2.

6.6 Linear Groups

In this final section we prove Theorem 6.3, which concerns a particular family of almost simple linear groups. Let $T = \mathrm{PSL}_n(q)$ where $n \geqslant 4$ is even and q is odd. We follow Sect. 6.4.2 very closely. Let us fix some notation.

Notation 6.6.1 Write $q = p^f$ where $f \geqslant 2$. Let $V = \mathbb{F}_q^n$.
Fix a basis $\mathcal{B} = (v_1, \dots, v_n)$ for V.
Fix the simple algebraic group $X = \mathrm{PSL}_n(\overline{\mathbb{F}}_p)$.
Fix the Frobenius endomorphism $\varphi = \varphi_{\mathcal{B}}$ and the standard graph automorphism $\gamma = \gamma_{\mathcal{B}}$ (see Definition 2.5.1).
Fix the antidiagonal element $\delta_2 = \delta^{\frac{q-1}{(q-1)_2}}$, where δ is given in Definition 6.2.1, so $|\delta_2| = (n, q-1)_2$ (see Remark 6.2.5).

In light of Remark 6.2.8, to prove Theorem 6.3, we can assume that $\theta = \delta_2\gamma\varphi^i$ where i divides f and $f/i \geqslant 3$ is odd.

Notation 6.6.1. (continued)
Write $q = q_0^e$ where $e = f/i$
Fix the Steinberg endomorphism $\sigma = \delta_2\gamma\varphi^i$ and the automorphism $\rho = \delta_2\gamma$.
Let $Z = C_X(\rho)^\circ$.

Proposition 6.6.2 *The automorphism ρ is an involution that commutes with σ and $Z_\sigma \cong \mathrm{PDO}^\eta_n(q_0)$ where $\eta = (-)^{\frac{n(q-1)}{4}+1}$.*

Proof Since e is odd, q_0-1 divides $q-1$ and $(q-1)/(q_0-1) = q_0^{e-1}+\cdots+q_0+1$ is odd, so $|\delta_2| = (n, q-1)_2 = (n, q_0-1)_2$. In particular, this implies that $\delta_2^{\varphi^i} = \delta_2$ and $\delta_2^\gamma = \delta_2^{-1}$. Therefore, $(\delta_2\gamma)^2 = 1$ and $(\delta_2\gamma\varphi^i)(\delta_2\gamma) = \gamma\varphi^i\gamma = (\delta_2\gamma)(\delta_2\gamma\varphi^i)$. Finally, by Remark 6.2.5(iv)

$$C_X(\rho)_\sigma = \{x \in X \mid x^{\delta_2\gamma\varphi^i} = x \text{ and } x^{\delta_2\gamma} = x\} = C_{\mathrm{PGL}_n(q_0)}(\delta_2\gamma) = \mathrm{PGO}^\eta_n(q_0),$$

and $Z_\sigma = \mathrm{PDO}^\eta_n(q_0)$. □

Proposition 6.6.3 *Let $T = \mathrm{PSL}_n(q)$ and $\theta = \delta_2\gamma\varphi^i$, where $n \geqslant 4$ is even and f/i is odd. Let $y \in \mathrm{PSO}^\eta_n(q_0) \leqslant T$ be the element in Fig. 6.8. Then there exists $t \in T$ that commutes with $\delta_2\gamma$ such that $(t\theta)^e$ is X-conjugate to $y\delta_2\gamma$.*

Proof From Proposition 6.6.2, we see $Z = C_X(\rho)^\circ$ is σ-stable. By Lemma 3.4.1(i), there exists $t \in \mathrm{PDO}^\eta_n(q_0) = Z_{\sigma^e} \leqslant X_{\gamma\sigma^e} = \mathrm{PGL}_n(q)$ such that $(t\sigma)^e$ is X-conjugate to $y\rho$. Moreover, since $y \in \mathrm{PSO}^\eta_n(q_0)$, by Lemma 5.4.3 we deduce that $t \in \mathrm{PSO}^\eta_n(q) \leqslant \mathrm{PSU}_n(q)$. □

Remark 6.6.4 This remark will help us understand how $y \in \mathrm{PDO}^\eta_n(q_0)$ from Fig. 6.9 acts on $V_0 = \mathbb{F}^n_{q_0^2}$ as an element of $\mathrm{PGU}_n(q_0)$.

We begin with some preliminaries, where we use [20, Proposition 3.3.2]. Let $g \in \mathrm{SO}^\varepsilon_{2d}(q_0)$ have type $(2d)^\varepsilon_{q_0}$ where $\varepsilon \in \{+,-\}$ and $d \geqslant 1$ (with d odd if $\varepsilon = -$). If $\varepsilon = -$ and d is even, then $|g| = \Phi^*_{2d}(q_0)$, and if $\varepsilon = +$ and d is odd, then

n	q	y
$n \leqslant 10$		$(n)^\eta_{q_0}$
$n \geqslant 12$ $0 \pmod 4$		$(n-2)^+_{q_0} \perp (2)^-_{q_0}$
$2 \pmod 4$	$1 \pmod 4$	$(\frac{n}{2}+1)^-_{q_0} \perp (\frac{n}{2}-3)^-_{q_0} \perp (2)^-_{q_0}$
	$3 \pmod 4$	$(n-2)^-_{q_0} \perp (2)^-_{q_0}$

Fig. 6.8 Linear groups: The element y

n	η	y	conditions
$n \leqslant 10$	$(-)^{\frac{n}{2}+1}$	$U \oplus U^*$	
	$(-)^{\frac{n}{2}}$	$W^1 \perp W^2$	
$n \geqslant 12$	$(-)^{\frac{n}{2}+1}$	$U \oplus U^* \perp W_0^1 \perp W_0^2$	
	$-$	$W_1^1 \perp W_1^2 \perp W_2^1 \perp W_2^2 \perp W_0^1 \perp W_0^2$	$n \equiv 2 \pmod 8$
	$-$	$(U_1 \oplus U_1^*) \perp (U_2 \oplus U_2^*) \perp W_0^1 \perp W_0^2$	$n \equiv 6 \pmod 8$

Note: $\dim U_1 = \dim W_1^j = (n+2)/4$ and $\dim U_2 = \dim W_2^j = (n-6)/4$, see Remark 6.6.4

Fig. 6.9 Linear groups: Decomposition centralised by y^2

$|g| = \Phi_d^*(q_0)$; in both cases, g centralises a decomposition $V_0 = U \oplus U^*$, where $\{U, U^*\}$ is a dual pair of totally singular d-spaces that are nonisomorphic irreducible $\mathbb{F}_{q_0^2}\langle g\rangle$-modules. Now assume that $\varepsilon = -$ and d is odd. Here $|g| = \Phi_{2d}^*(q_0)$ and $2d \equiv 2 \pmod 4$, so g centralises a decomposition $V_0 = U_1 \oplus U_2$, where U_1 and U_2 are nondegenerate d-spaces that are nonisomorphic irreducible $\mathbb{F}_{q_0^2}\langle g\rangle$-modules.

This allows us to obtain a decomposition of V_0 centralised by y^2, which we present in Fig. 6.9. Let us explain our notation. For any symbol S, the subspaces S^1 and S^2 are equidimensional. The subspaces W^j and W_i^j are nondegenerate and the subspaces U and U_i are totally singular. In every decomposition, the summands are pairwise nonisomorphic irreducible $\mathbb{F}_{q_0^2}\langle g\rangle$-modules, except when $n \geqslant 12$ and $q_0 = 3$, where y^2 acts as $-I_2$ on $W_0^1 \perp W_0^2$.

Proposition 6.6.5 *Assume that $n \geqslant 12$. Then the maximal subgroups of G that contain $t\theta$ are listed in Fig. 6.10, where $m(H)$ is an upper bound on the multiplicity of the subgroups of type H in $\mathcal{M}(G, t\theta)$.*

Proof First assume that $H \in \mathcal{C}_1$. It is straightforward to determine the maximal reducible subgroups of $\mathrm{PGU}_n(q_0)$ that contain y^2 by using Remark 6.6.4 (if $n \equiv 2 \pmod 4$ and $\eta = -$, then there are several but we simply note that y^2 is contained in at most 6 of any given type). Lemma 3.4.1(ii)(b) now implies that the $\mathcal{C}_1$ subgroups of G that contain $t\theta$ are the corresponding subgroups that appear in Fig. 6.10 (see the proof of Proposition 6.4.10 for further details).

Now assume that $H \notin \mathcal{C}_1$. An X-conjugate of $(t\theta)^{2e}$ is $(y\gamma)^2 = y^2$. If $q_0 > 3$, then a suitable power z of y^2 has type $[\lambda, \lambda^{q_0}] \perp I_{n-2}$ where $\lambda \in \mathbb{F}_{q_0^2}^{\times}$ satisfies

	type of H	$m(H)$	conditions
$\mathcal{C}_1$	$\mathrm{GL}_1(q) \times \mathrm{GL}_{n-1}(q)$	2	
	$P_{1,n-1}$	2	$q_0 = 3$
	$\mathrm{GL}_2(q) \times \mathrm{GL}_{n-2}(q)$	1	
	$P_{(n-2)/2,(n+2)/2}$	2	$\eta = (-)^{\frac{n}{2}+1}$
	$P_{k,n-k}$	6	$n \equiv 2 \pmod 4$, $\eta = -$, $1 < k < n/2$
	$\mathrm{GL}_k(q) \times \mathrm{GL}_{n-k}(q)$	6	$n \equiv 2 \pmod 4$, $\eta = -$, $1 < k < n/2$
$\mathcal{C}_2$	$\mathrm{GL}_{n/k}(q) \wr S_k$	N	$k > 1$, $k \mid n$
$\mathcal{C}_3$	$\mathrm{GL}_{n/2}(q^2)$	N	
$\mathcal{C}_4$	$\mathrm{GL}_2(q) \otimes \mathrm{GL}_{n/2}(q)$	N	$q_0 = 3$
$\mathcal{C}_5$	$\mathrm{GL}_n(q^{1/k})$	N	k prime, $k \mid f$
$\mathcal{C}_8$	$\mathrm{Sp}_n(q)$	N	
	$\mathrm{O}_n^{\upsilon}(q)$	N	$\upsilon \in \{+, -\}$
	$\mathrm{GU}_n(q^{1/2})$	N	f even

Note: $N = |C_{\mathrm{PGU}_n(q_0)}(y^2)|$

Fig. 6.10 Linear groups: Description of $\mathcal{M}(G, t\theta)$

$\lambda \neq \lambda^{q_0}$, and if $q_0 = 3$, then a power z of y^2 is $-I_2 \perp I_{n-2}$. In both cases $\nu(z) = 2$ and Proposition 4.2.5 implies that H appears in Fig. 6.6. Since geometric maximal subgroups of G of a given type are $\langle \mathrm{PGL}_n(q), \theta\rangle$-conjugate, Lemma 3.4.1(ii)(a) gives $m(H)$, noting that Proposition 6.4.7 implies $(\rho\sigma^e)^2 = \sigma^{2e}$. □

Proof of Theorem 6.3 As before, we apply Lemma 2.1.1. Let $x \in G$ have prime order. By [14, Corollary 1], if $n \geqslant 8$ and $H \leqslant G$ is a maximal nonsubspace subgroup, then

$$\mathrm{fpr}(x, G/H) < \frac{2}{q^{n-3}},$$

and by [21, Theorem 2.7], if H has type $\mathrm{GL}_k(q) \times \mathrm{GL}_{n-k}(q)$ or $P_{k,n-k}$, with $k < n/2$, then

$$\mathrm{fpr}(x, G/H) \leqslant \begin{cases} q^{-1} + q^{-(n-1)} & \text{if } k = 1 \\ 2q^{-k} & \text{if } k > 1. \end{cases}$$

First assume that $n \geqslant 12$ and $\eta = (-)^{\frac{n}{2}+1}$. From Remark 6.6.4, we see

$$|C_{\mathrm{PGU}_n(q_0)}(y^2)| = \begin{cases} (3^2 - 1)(3^{n-1} - 3) & \text{if } q_0 = 3 \\ (q_0 + 1)(q_0^{n-2} - 1) & \text{otherwise.} \end{cases}$$

Therefore, writing $d(n)$ for the number of divisors of n, we have

$$P(x, t\theta) < (d(n) + \log\log q + 4) \cdot (q_0^2 - 1)(q_0^{n-1} - q_0) \cdot \frac{2}{q^{n-3}} + \frac{1}{q} + \frac{1}{q^{n-1}} + \frac{2}{q^2} + \frac{4}{q^{(n-2)/2}} < \frac{1}{2}.$$

Next assume that $\eta = -$ and $n \geqslant 14$ satisfies $n \equiv 2 \pmod 4$. Then

$$|C_{\mathrm{PGU}_n(q_0)}(y^2)| = \begin{cases} (q_0 + 1)(q_0^{(n+2)/4} + 1)^2(q_0^{(n-6)/4} + 1)^2 & \text{if } n \equiv 2 \pmod 8 \\ (q_0 + 1)(q_0^{(n+2)/2} - 1)(q^{(n-6)/2} - 1) & \text{if } n \equiv 6 \pmod 8). \end{cases}$$

Therefore,

$$P(x, t\theta) < (d(n) + \log\log q + 4) \cdot (q_0 + 1)(q_0^{(n+2)/4} + 1)^2(q_0^{(n-6)/2} + 1)^2 \cdot \frac{2}{q^{n-3}} + \frac{1}{q} + \frac{1}{q^{n-1}} + \frac{2}{q^2} + 6\sum_{k\geqslant 2} \frac{2}{q^k} < \frac{1}{2}.$$

Now assume that $n \in \{6, 8, 10\}$. Arguing as in the proof of Proposition 6.6.5, via Lemma 3.4.1(ii)(b), $t\theta$ is not contained in any reducible maximal subgroups of G.

From [8] we see that G has at most $13 + \log\log q$ classes of irreducible maximal subgroups. Note that

$$|C_{\mathrm{PGU}_n(q_0)}(y^2)| \leqslant \frac{(q_0^{n/2}+1)^2}{q_0+1}.$$

Using the fixed point ratio bound from [21, Corollary 2.9], we obtain

$$P(x, t\theta) < (13 + \log\log q) \cdot (q_0^{n/2}+1) \cdot \left(\frac{q-1}{(q^{n-1}-1)(q^n-1)}\right)^{1/2-1/n} < \frac{1}{2}.$$

Finally assume that $n = 4$. As in the previous cases, $t\theta$ is not contained in any reducible maximal subgroups of G, there are at most $6 + \log\log q$ classes of irreducible maximal subgroups and $|C_{\mathrm{PGU}_4(q_0)}(y^2)| = (q_0-1)(q_0^2+1)$. As with the four-dimensional unitary groups, the subgroups of type $\mathrm{Sp}_4(q)$ present a special challenge. If $H \in \mathcal{M}(G, t\theta)$ does not have type $\mathrm{Sp}_4(q)$, then [21, Corollary 2.9] implies that

$$\mathrm{fpr}(x, G/H) < \left((q+1)(q^2+1)(q^3-1)^2\right)^{-1/4},$$

and if H has type $\mathrm{Sp}_4(q)$, then [21, Lemma 2.11] gives

$$\mathrm{fpr}(x, G/H) < \frac{q^2}{(2, q-1)(q^3-1)}.$$

Therefore,

$$P(x, t\theta) < \frac{(6+\log\log q)\cdot(q_0-1)(q_0^2+1)}{\left((q+1)(q^2+1)(q^3-1)^2\right)^{-1/4}} + \frac{(q_0-1)(q_0^2+1)\cdot q^2}{(2, q-1)(q^3-1)} < \frac{1}{2}.$$

As usual, in all cases $P(x, t\theta) \to 0$ as $q \to \infty$. □

It remains to note that Theorem 4 is a combination of Theorems 5.1 and 6.1, and similarly Theorem 5 is a combination of Theorems 5.2 and Theorem 6.2. Moreover, Theorems 2 and 3 follow from Theorems 4 and 5, together with the relevant results on linear groups in [21] and Theorem 6.3 and the relevant results on symplectic and odd-dimensional orthogonal groups in [43]. Theorem 1 is a corollary of Theorem 2 (noting that $s(S_6) = 2$).

Appendix A
MAGMA Code

In this appendix, we give the MAGMA [6] code for our computational methods. See Sect. 2.7 for further information.

The function `FixedPointRatio` calculates the fixed point ratio $\mathrm{fpr}(g, G/H)$ of an element $g \in G$ in the action of G on G/H. It takes as input a group G, a subgroup $H \leqslant G$ and an element $g \in G$. The function returns the fixed point ratio $\mathrm{fpr}(g, G/H)$.

```
function FixedPointRatio( G, H, g )
  count:=0;
  classreps:=Classes(H);
  for rep in classreps do
    if (rep[1] eq Order(g)) then
      if IsConjugate(G,g,rep[3]) then
        count:=count+rep[2];
      end if;
    end if;
  end for;
  return count*Order(Centraliser(G,g))/Order(G);
end function;
```

The function `MaximalOvergroups` provides information about the maximal overgroups of an element. The input is a group G and an element $s \in G$. The function returns a pair of lists $[H_1, \ldots, H_m]$ and $[k_1, \ldots, k_m]$ where H_i are pairwise non-conjugate maximal subgroups of G and k_i is the number of conjugates of H_i which contain s.

```
function MaximalOvergroups( G, s )
  groups:=[];
  mults:=[];
  maxes:=MaximalSubgroups(G : OrderMultipleOf:=Order(s));
  for M in maxes do
    H:=M`subgroup;
    count:=FixedPointRatio(G,H,s)*Order(G)/Order(H);
    if (count ne 0) then
```


S. Harper, *The Spread of Almost Simple Classical Groups*, Lecture Notes in Mathematics 2286, https://doi.org/10.1007/978-3-030-74100-6

```
      groups:=Append(groups,H);
      mults:=Append(mults,count);
    end if;
  end for;
  return <groups, mults>;
end function;
```

The function `ClassRepTuples` is based heavily on an algorithm of Breuer [10, Section 3.3]. The input is a group G and a list $[x_1, \ldots, x_k]$ of elements of G. The function returns a list of orbit representatives for the diagonal conjugation action of G on $x_1^G \times \cdots \times x_k^G$.

```
function ClassRepTuples( G, list )
  cents:=[];
  for x in list do
    cents:=Append(cents,Centraliser(G,x));
  end for;
  function OrbReps(G, reps, int, i, cents, list )
    if (i gt #list) then
      L:=[reps];
    else
      L:=[];
      for r in DoubleCosetRepresentatives(G,cents[i],int) do
        L:=L cat OrbReps(G, Append(reps,list[i]^r),
          (int meet cents[i]^r), i+1, cents, list );
      end for;
    end if;
    return L;
  end function;
  return OrbReps(G,[list[1]],cents[1],2,cents,list);
end function;
```

The function `RandomCheck` is a randomised algorithm that plays a role in determining the uniform spread of a group. The input is a group G, an element $s \in G$, a list $[x_1, \ldots, x_k]$ of elements in G and a nonnegative integer N. The claim to be tested is: for every list $[y_1, \ldots, y_k]$ with $y_i \in x_i^G$, there exists $z \in s^G$ such that $\langle y_1, z\rangle = \cdots = \langle y_k, z\rangle = G$. If the function returns `true`, then this claim is true, and if the function returns `false`, then the result is inconclusive. The claim is tested by random selections of elements in G, the number of which depends on the parameter N.

```
function RandomCheck( G, s, list, N )
  classtuples:=ClassRepTuples(G,list);
  for X in classtuples do
    found:=false;
    for i in [1..N] do
      h:=Random(G);
      found:=true;
      for x in X do
        H:=sub<G|[x,s^h]>;
        if not (Order(H) eq Order(G)) then
          found:=false;
```

```
          break;
        end if;
      end for;
      if (found) then
        break;
      end if;
    end for;
    if (not found) then
      return false;
    end if;
  end for;
  return true;
end function;
```

The function ProbabilisticMethod is our main computational tool for studying the uniform spread of a group. The input is a group G, an element $s \in G$ and nonnegative integers k and N. First, the function implements the probabilistic method described in Sect. 2.1 to determine whether $u(G) \geqslant k$ with respect to the class s^G. If successful, the function returns true; otherwise the second phase commences. Here RandomCheck is employed to verify that for all $(y_1, \ldots, y_k)$ with $y_i \in x_i^G$ there exists $z \in s^G$ such that $\langle y_1, z \rangle = \cdots = \langle y_k, z \rangle$, for all k-tuples $(x_1^G, \ldots, x_k^G)$ of conjugacy classes for which this was not proved in the first phase. If successful, the function returns true. If false is returned, then the result is inconclusive. A variety of helpful data from the computation is printed to the standard output.

```
function ProbabilisticMethod( G, s, k, N )
  maxandmult:=MaximalOvergroups(G,s);
  max:=maxandmult[1];
  mult:=maxandmult[2];

  print "--- \nMAXIMAL SUBGROUPS \n--- \n ";
  for i in [1..#max] do
    print [Order(max[i]), mult[i]];
  end for;
  print " ";

  classes:=Classes(G);
  primeclasses:=[];
  sums:=[];

  print "--- \nCONJUGACY CLASSES \n--- \n ";
  for class in classes do
    if (IsPrime(class[1])) then
      primeclasses:=Append(primeclasses,class[3]);
      ratios:=[];
      for H in max do
        ratios:=Append(ratios,FixedPointRatio(G,H,class[3]));
      end for;
      sum:=0;
      for i in [1..#max] do
        sum:=sum+ratios[i]*mult[i];
```

```
    end for;
    sums:=Append(sums,sum);
    print "Order:", class[1];
    print "Size:", class[2];
    print "Fixed Point Ratios:", ratios;
    print "Sum of FPRs:", sum;
    print " \n--------------\n ";
  end if;
end for;

print "--- \nBAD TUPLES \n--- \n ";

tuples:=[];
if exists{sum: sum in sums | sum ge 1/k} then
  markers:=[1 .. #sums];
  ind:=[[]];
  for i in [1 .. k] do
    newind:=[];
    for y in ind do
      for x in markers do
        if (i eq 1) or (x ge y[i-1]) then
          z:=Append(y,x);
          newind:=Append(newind,z);
        end if;
      end for;
    end for;
    ind:=newind;
  end for;
  seq:=[];
  for I in ind do
    elt:=[];
    for i in I do
      elt:=Append(elt,sums[i]);
    end for;
    seq:=Append(seq,elt);
  end for;
  for i in [1 .. #seq] do
    tot:=0;
    for x in seq[i] do
      tot:=tot+x;
    end for;
    if tot ge 1 then
      tuples:=Append(tuples,ind[i]);
    end if;
  end for;
end if;

print "Bad Tuples:", tuples;
print " ";
if N gt 0 then
  badtuples:=[];
  for tuple in tuples do
    list:=[];
    for t in tuple do
```

```
        list:=Append(list, primeclasses[t]);
      end for;
      if not RandomCheck(G,s,list,N) then
        badtuples:=Append(badtuples,tuple);
      end if;
    end for;
    print "Bad tuples remaining after", N,
      "random checks:", badtuples;
    print " ";
  else
    badtuples:=tuples;
  end if;

  return (badtuples eq []);
end function;
```

We sometimes want to work with groups that cannot be handled with the command `MaximalSubgroups`. In this case, we use the function `ClassicalMaximals`. For example, to obtain the maximal subgroups of $O_{12}^{+}(2)$ we use

```
ClassicalMaximals("O+", 12, 2 : general:=true);
```

References

1. M. Aschbacher, On the maximal subgroups of the finite classical groups. Invent. Math. **76**, 469–514 (1984)
2. M. Aschbacher, *Finite Group Theory*. Cambridge Studies in Advanced Mathematics, vol. 10, 2nd ed. (Cambridge University Press, Cambridge, 2000)
3. M. Aschbacher, R. Guralnick, Some applications of the first cohomology group. J. Algebra **90**, 446–460 (1984)
4. M. Aschbacher, G.M. Seitz, Involutions in Chevalley groups over fields of even order. Nagoya Math. J. **63**, 1–91 (1976)
5. J. Bamberg, T. Penttila, Overgroups of cyclic Sylow subgroups of linear groups. Commun. Algebra **36**, 2503–2543 (2008)
6. W. Bosma, J. Cannon, C. Playoust, The MAGMA algebra system I: the user language. J. Symbolic Comput. **24**, 235–265 (1997)
7. J.N. Bray, D.F. Holt, C.M. Roney-Dougal, Certain classical groups are not well-defined. J. Group Theory **12**, 171–180 (2009)
8. J.N. Bray, D.F. Holt, C.M. Roney-Dougal, *The Maximal Subgroups of the Low-Dimensional Finite Classical Groups*. London Math. Soc. Lecture Notes Series, vol. 407 (Cambridge University Press, Cambridge, 2013)
9. J.L. Brenner, J. Wiegold, Two generator groups, I. Michigan Math. J. **22**, 53–64 (1975)
10. T. Breuer, GAP computations concerning probabilistic generation of finite simple groups (2007). arXiv:0710.3267
11. T. Breuer, R.M. Guralnick, W.M. Kantor, Probabilistic generation of finite simple groups, II. J. Algebra **320**, 443–494 (2008)
12. T. Breuer, R.M. Guralnick, A. Lucchini, A. Maróti, G.P. Nagy, Hamiltonian cycles in the generating graphs of finite groups. Bull. Lond. Math. Soc. **42**, 621–633 (2010)
13. T. Brookfield, *Overgroups of a linear Singer cycle in classical groups*, Ph.D. Thesis, University of Birmingham, 2014
14. T.C. Burness, Fixed point ratios in actions of finite classical groups, I. J. Algebra **309**, 69–79 (2007)
15. T.C. Burness, Fixed point ratios in actions of finite classical groups, II. J. Algebra **309**, 80–138 (2007)
16. T.C. Burness, Fixed point ratios in actions of finite classical groups, III. J. Algebra **314**, 693–748 (2007)
17. T.C. Burness, Fixed point ratios in actions of finite classical groups, IV. J. Algebra **314**, 749–788 (2007)

S. Harper, *The Spread of Almost Simple Classical Groups*, Lecture Notes in Mathematics 2286, https://doi.org/10.1007/978-3-030-74100-6

18. T.C. Burness, Simple groups, fixed point ratios and applications, in *Local Representation Theory and Simple Groups*. EMS Series of Lectures in Mathematics (European Mathematical Society, 2018), pp. 267–322
19. T.C. Burness, Simple groups, generation and probabilistic methods, in *Proceedings of Groups St Andrews 2017*. London Math. Soc. Lecture Note Series, vol. 455 (Cambridge University Press, Cambridge, 2019), pp. 200–229
20. T.C. Burness, M. Giudici, *Classical Groups, Derangements and Primes*. Aust. Math. Soc. Lecture Note Series, vol. 25 (Cambridge University Press, Cambridge, 2016)
21. T.C. Burness, S. Guest, On the uniform spread of almost simple linear groups. Nagoya Math. J. **209**, 35–109 (2013)
22. T.C. Burness, R.M. Guralnick, S. Harper, The spread of a finite group. Ann. of Math. **193**, 619–687 (2021)
23. T.C. Burness, S. Harper, On the uniform domination number of a finite simple group. Trans. Am. Math. Soc. **372**, 545–583 (2019)
24. T.C. Burness, S. Harper, Finite groups, 2-generation and the uniform domination number. Israel J. Math. **239**, 271–367 (2020)
25. A.A. Buturlakin, M.A. Grechkoseeva, The cyclic structure of maximal tori of the finite classical groups. Algebra Logic **46**, 73–89 (2007)
26. M. Cabanes, B. Späth, Descent equalities and the inductive McKay condition for types B and E. Adv. Math. **356**, 106820 (2019)
27. R.W. Carter, *Simple Groups of Lie Type* (Wiley, London, 1972)
28. B. Chang, The conjugate classes of Chevalley groups of type (G_2). J. Algebra **9**, 190–211 (1968)
29. F. Digne, J. Michel, Groupes réductifs non connexes. Ann. Sci. École Norm. Sup. (4) **27**, 345–406 (1994)
30. C. Donoven, S. Harper, Infinite $\frac{3}{2}$-generated groups. Bull. Lond. Math. Soc. **52**, 657–673 (2020)
31. H. Enomoto, The conjugacy classes of Chevalley groups of type (G_2) over finite fields of characteristic 2 or 3. J. Fac. Sci. Univ. Tokyo Sect. I **16**, 497–512 (1969)
32. F. Erdem, On the generating graphs of symmetric groups. J. Group Theory **21**, 629–649 (2018)
33. B. Fairbairn, The exact spread of M_{23} is 8064. Int. J. Group Theory **1**, 1–2 (2012)
34. D. Frohardt, K. Magaard, Grassmannian fixed point ratios. Geom. Dedicata **82**, 21–104 (2000)
35. D. Frohardt, K. Magaard, Composition factors of monodromy groups. Ann. of Math. **154**, 327–345 (2001)
36. D. Gorenstein, R. Lyons, *The Local Structure of Finite Groups of Characteristic 2 Type*, vol. 276 (Mem. Amer. Math. Soc., 1983)
37. D. Gorenstein, R. Lyons, R. Solomon, *The Classification of the Finite Simple Groups, Number 3*, Mathematical Surveys and Monographs, vol. 40 (Amer. Math. Soc., 1998)
38. R.M. Guralnick, W.M. Kantor, Probabilistic generation of finite simple groups. J. Algebra **234**, 743–792 (2000)
39. R.M. Guralnick, G. Malle, Products of conjugacy classes and fixed point spaces. J. Am. Math. Soc. **25**, 77–121 (2012)
40. R.M. Guralnick, T. Penttila, C.E. Praeger, J. Saxl, Linear groups with orders having certain large prime divisors. Proc. Lond. Math. Soc. **78**, 167–214 (1997)
41. R.M. Guralnick, J. Saxl, Generation of finite almost simple groups by conjugates. J. Algebra **268**, 519–571 (2003)
42. R.M. Guralnick, A. Shalev, On the spread of finite simple groups. Combinatorica **23**, 73–87 (2003)
43. S. Harper, On the uniform spread of almost simple symplectic and orthogonal groups. J. Algebra **490**, 330–371 (2017)
44. S. Harper, Shintani descent, simple groups and spread. J. Algebra **578**, 319–355 (2021)
45. N. Kawanaka, On the irreducible characters of the finite unitary groups. J. Math. Soc. Jpn. **29**, 425–450 (1977)

46. R. Kessar, Shintani descent and perfect isometries for blocks of finite general linear groups. J. Algebra **276**, 493–501 (2004)
47. P.B. Kleidman, The maximal subgroups of the finite 8-dimensional orthogonal groups $P\Omega_8^+(q)$ and of their automorphism groups. J. Algebra **110**, 173–242 (1987)
48. P.B. Kleidman, The maximal subgroups of the Chevalley groups $G_2(q)$ with q odd, the Ree groups ${}^2G_2(q)$ and their automorphism groups. J. Algebra **117**, 30–71 (1988)
49. P.B. Kleidman, The maximal subgroups of the Steinberg triality groups ${}^3D_4(q)$ and of their automorphism groups. J. Algebra **115**, 182–199 (1988)
50. P.B. Kleidman, M.W. Liebeck, *The Subgroup Structure of the Finite Classical Groups*, London Math. Soc. Lecture Note Series, vol. 129 (Cambridge University Press, Cambridge, 1990)
51. S. Lang, *Algebra*. Graduate Texts in Mathematics, vol. 211, 3rd edn. (Springer, Berlin, 2002)
52. R. Lawther, M.W. Liebeck, G.M. Seitz, Fixed point ratios in actions of finite exceptional groups of Lie type. Pac. J. Math. **205**, 393–463 (2002)
53. M.W. Liebeck, Probabilistic and asymptotic aspects of finite simple groups, in *Probabilistic Group Theory, Combinatorics, and Computing*. Lecture Notes in Math., vol. 2070 (Springer, Berlin, 2013), pp. 1–34
54. M.W. Liebeck, J. Saxl, Minimal degrees of primitive permutation groups, with an application to monodromy groups of covers of Riemann surfaces. Proc. Lond. Math. Soc. **63**, 266–314 (1991)
55. M.W. Liebeck, A. Shalev, Simple groups, permutation groups, and probability. J. Am. Math. Soc. **12**, 497–520 (1999)
56. S. Piccard, Sur les bases du groupe symétrique et du groupe alternant. Math. Ann. **116**, 752–767 (1939)
57. A. Shalev, Probabilistic group theory and Fuchsian groups, in *Infinite Groups: Geometric, Combinatorial and Dynamical Aspects*. Progr. Math., vol. 248 (Birkhäuser, Basel, 2005), pp. 363–388
58. T. Shintani, Two remarks on irreducible characters of finite general linear groups. J. Math. Soc. Jpn. **28**, 396–414 (1976)
59. T. Shoji, Character sheaves and almost characters of reductive groups. Adv. Math. **111**, 244–313 (1995)
60. R. Steinberg, *Endomorphisms of Linear Algebraic Groups*, vol. 80 (Mem. Amer. Math. Soc., 1968)
61. R. Steinberg, Generators for simple groups. Can. J. Math. **14**, 277–283 (1962)
62. A.J. Woldar, $\frac{3}{2}$-generation of the sporadic simple groups. Commun. Algebra **22**, 675–685 (1994)
63. K. Zsigmondy, Zur Theorie der Potenzreste. Monat. Math. Phys. **3**, 265–284 (1892)

LECTURE NOTES IN MATHEMATICS Springer

Editorial Policy

1. Lecture Notes aim to report new developments in all areas of mathematics and their applications – quickly, informally and at a high level. Mathematical texts analysing new developments in modelling and numerical simulation are welcome.

   Manuscripts should be reasonably self-contained and rounded off. Thus they may, and often will, present not only results of the author but also related work by other people. They may be based on specialised lecture courses. Furthermore, the manuscripts should provide sufficient motivation, examples and applications. This clearly distinguishes Lecture Notes from journal articles or technical reports which normally are very concise. Articles intended for a journal but too long to be accepted by most journals, usually do not have this "lecture notes" character. For similar reasons it is unusual for doctoral theses to be accepted for the Lecture Notes series, though habilitation theses may be appropriate.

2. Besides monographs, multi-author manuscripts resulting from SUMMER SCHOOLS or similar INTENSIVE COURSES are welcome, provided their objective was held to present an active mathematical topic to an audience at the beginning or intermediate graduate level (a list of participants should be provided).

   The resulting manuscript should not be just a collection of course notes, but should require advance planning and coordination among the main lecturers. The subject matter should dictate the structure of the book. This structure should be motivated and explained in a scientific introduction, and the notation, references, index and formulation of results should be, if possible, unified by the editors. Each contribution should have an abstract and an introduction referring to the other contributions. In other words, more preparatory work must go into a multi-authored volume than simply assembling a disparate collection of papers, communicated at the event.

3. Manuscripts should be submitted either online at www.editorialmanager.com/lnm to Springer's mathematics editorial in Heidelberg, or electronically to one of the series editors. Authors should be aware that incomplete or insufficiently close-to-final manuscripts almost always result in longer refereeing times and nevertheless unclear referees' recommendations, making further refereeing of a final draft necessary. The strict minimum amount of material that will be considered should include a detailed outline describing the planned contents of each chapter, a bibliography and several sample chapters. Parallel submission of a manuscript to another publisher while under consideration for LNM is not acceptable and can lead to rejection.

4. In general, **monographs** will be sent out to at least 2 external referees for evaluation.

   A final decision to publish can be made only on the basis of the complete manuscript, however a refereeing process leading to a preliminary decision can be based on a pre-final or incomplete manuscript.

   Volume Editors of **multi-author works** are expected to arrange for the refereeing, to the usual scientific standards, of the individual contributions. If the resulting reports can be forwarded to the LNM Editorial Board, this is very helpful. If no reports are forwarded or if other questions remain unclear in respect of homogeneity etc, the series editors may wish to consult external referees for an overall evaluation of the volume.

5. Manuscripts should in general be submitted in English. Final manuscripts should contain at least 100 pages of mathematical text and should always include

   - a table of contents;
   - an informative introduction, with adequate motivation and perhaps some historical remarks: it should be accessible to a reader not intimately familiar with the topic treated;
   - a subject index: as a rule this is genuinely helpful for the reader.
   - For evaluation purposes, manuscripts should be submitted as pdf files.

6. Careful preparation of the manuscripts will help keep production time short besides ensuring satisfactory appearance of the finished book in print and online. After acceptance of the manuscript authors will be asked to prepare the final LaTeX source files (see LaTeX templates online: https://www.springer.com/gb/authors-editors/book-authors-editors/manuscriptpreparation/5636) plus the corresponding pdf- or zipped ps-file. The LaTeX source files are essential for producing the full-text online version of the book, see http://link.springer.com/bookseries/304 for the existing online volumes of LNM). The technical production of a Lecture Notes volume takes approximately 12 weeks. Additional instructions, if necessary, are available on request from lnm@springer.com.

7. Authors receive a total of 30 free copies of their volume and free access to their book on SpringerLink, but no royalties. They are entitled to a discount of 33.3 % on the price of Springer books purchased for their personal use, if ordering directly from Springer.

8. Commitment to publish is made by a *Publishing Agreement*; contributing authors of multiauthor books are requested to sign a *Consent to Publish form*. Springer-Verlag registers the copyright for each volume. Authors are free to reuse material contained in their LNM volumes in later publications: a brief written (or e-mail) request for formal permission is sufficient.

Addresses:

Professor Jean-Michel Morel, CMLA, École Normale Supérieure de Cachan, France
E-mail: moreljeanmichel@gmail.com

Professor Bernard Teissier, Equipe Géométrie et Dynamique,
Institut de Mathématiques de Jussieu – Paris Rive Gauche, Paris, France
E-mail: bernard.teissier@imj-prg.fr

Springer: Ute McCrory, Mathematics, Heidelberg, Germany,
E-mail: lnm@springer.com

Printed in the United States
by Baker & Taylor Publisher Services